ACTIVE Biology

ACTIVE Biology

MIKE BAILEY, KEITH HIRST AND RAY WATKIN

HODDER AND STOUGHTON
LONDON SYDNEY AUCKLAND TORONTO

Acknowledgements

The authors and publishers would like to thank the following for their permission to reproduce copyright material:
Burke Publishing Company Ltd for the drawings on page 13 from *the Young Specialist Looks at Pond Life* by W Engelhardt; HMSO and the Forestry Commission for 'Conifer Seeds' on page 14 from *Know your Conifers* by L Edlin Crown copyright reproduced by permission of the Forestry Commission; Armand Colin for the figure on page 22 from *Ecologie Classe de Seconde* by Boden, Caro, Lesec and Pedotti; Heinemann Educational Books Ltd for the map and data on page 27 from *Human Ecology* by B Campbell; Basil Blackwell for the figures on page 32 and the data on page 28 from *Microbes and Biotechnology* by M R Ingle; The Association for Consumer Research for the articles on page 40 from *Gardening Which* April 1985 and May 1986; Chapman and Hall for the data on page 43 from *The Living Body* by C Best and Taylor; Churchill Livingstone for the data on page 44 and the table on page 58 and the table on page 95 from *Textbook of Physiology* by Bell *et al* and the figure on page 86 from *Clinical Chemistry in Practical Medicine* by Stewart and Dunlop; Oxford University Press for the figure on page 51 from *An Introduction to Plant Physiology* by W O James © Oxford University Press 1973; Edward Arnold for the figure on page 52 from *Photosynthesis* by D O Hall and K K Rao; Unwin and Hyman for tables on page 54 and page 83 from *Advanced Biology* by Simpkin and Williams published by Bell & Hyman; Harper & Row for the figure on page 55 from *The Experimental analysis of distribution and abundance* by C J Krebs; Cambridge University Press for the material on page 63 from *Human Biology* by P Rowlinson and M Jenkins and the table on page 104 from *Biological Science vol 2* by Green *et al*; John Murray for the table on page 74 from *Problems in Animal Physiology* by M K Sands; Cambridge University Press for the table on page 83 from *The Nature of Biochemistry* by Baldwin; McGraw Hill for the material on page 101 from *Principles of Genetics* by Sinnott, Dunn, Dobjhansky; The Institute of Biology for the material on page 77 from 'Exercise in Birds: a study using radiotelemetry' by P J Butler published in *The Biologist: Journal of the Institute of Biology* vol 32, no. 2; and the material on page 105 from 'Wheat by F G H Lupton published in *Biologist: Journal of the Institute of Biology* vol 32, no. 2, 1985 and the table on page 61 from 'Pigs' by J W B King from *Biologist: Journal of the Institute of Biology* vol 30 no. 5; Edward Arnold for the material on page 36 from *Science Through Biology* by J J Head; Blackie and Son for the table on page 37 from *Biology in Action* by David Luxton; The Hamlyn Publishing Group for the table on page 96 from *La Rousse Encyclopaedia of Animal Life*.

They are also grateful to the following for permission to use their photographs:
The Bettman Archive, Inc page 60; Biofotos, pages 1, 2 (top), 3, 4, 40; K G Brocklehurst, page 42 (top right and left, bottom left); Central Office of Information, page 67; A W Curtis, page 74; Forest Films, page 2 (bottom); Dr L H Jones (in *Biologist* **30** (4) 1983), page 99 (top); Dr Alan McDermott (in *Biologist* **27** (3) 1980), page 93; Don Mackean, page 99 (top); Mars Confectionary, page 57; The Metropolitan Police, page 97; Planet Earth Pictures, pages 78, 81; Science Photo Library, pages 42 (bottom right), 66, 69, 91.

Every effort has been made to trace copyright holders of material reproduced in this book. Any rights not acknowledged here will be acknowledged in subsequent printings if notice is given to the publisher.

First published in Great Britain in 1989

British Library Cataloguing in Publication Data
Bailey, Mike
Active Biology.
Pupil's book
1. Biology – For schools
I. Title II. Hirst, Keith III. Watkin, Ray
574
ISBN 0 7131 7756 X

Typeset in Rockwell by Tradespools Ltd, Frome, Somerset.
Printed in Great Britain for the educational publishing division of Hodder and Stoughton Ltd, Mill Road, Dunton Green, Sevenoaks, Kent by Thomson Litho Ltd, East Kilbride, Glasgow.

Contents

To the Student

This book is intended to help you practise the skills that you will be developing in your Biology or Science GCSE course. The questions do not simply test whether you can remember the facts from your course, important though that is. Many of them test your understanding of basic principles, and often you will need to apply your knowledge and understanding of basic principles to solve problems. The questions may often be about topics which you have not studied in your course and some may look difficult at first sight. Indeed some are difficult! We do not expect you to try all the questions — you will need to select the ones that are suitable for you. However, we hope that you will think about what you have learnt and use your knowledge to answer unfamiliar questions. In this way you will improve your ability to tackle the novel questions that you will get in your GCSE exams, and at the same time you will extend your knowledge and understanding of Biology.

As well as teaching you about living organisms, your course should be helping you to develop a variety of other useful skills. Many of the tasks in this book will assist further in this process, as long as you go about them with care and thought. Some of these skills are:

1. careful observation;
2. accurate measuring, and counting of large numbers in a systematic way;
3. drawing simple, clear diagrams;
4. accurately extracting data, for example from tables or in reading graphs;
5. handling data, e.g. carrying out simple calculations of percentages using the results of your experiments, or drawing graphs and pie charts;
6. interpreting data, for instance in making inferences from observations and drawing conclusions from experimental results;
7. suggesting sensible hypotheses, based on observations, which can be tested experimentally;
8. designing and planning experiments;
9. understanding the principles on which a scientific experiment is based, e.g. controlling variables;
10. understanding the economic and technological principles which relate to Biological Science.

You will find examples of questions which will give you practice in all these. We will not delude ourselves that you will look forward with great joy to doing all of the questions, but we hope that you will find at least some of them interesting and that they will stimulate you to find our more about the living world. We also hope that they will help you to achieve success in your GCSE exams.

Good Luck!

Organisms and their habitats

Living organisms are found in a great variety of different places, from the Arctic snows to tropical forests, and from hot dry deserts to the bottom of deep oceans. However, each type of living organism can usually live only in one kind of place, called its habitat. For example, polar bears live in the Arctic snows, but not in tropical forests; giant squids live in deep oceans but never in deserts! The questions in this unit are about habitats, and about how the conditions in a habitat affect the living organisms.

1 Habitats

a The photographs below show eight different living organisms. Match each organism with the habitat in which you would be most likely to find it.

Choose from this list of habitats.

freshwater pond
garden soil
grassy field
moorland
mountain stream
seashore
stone wall
woodland floor

bluebell

daisy

crab

earthworm

heather

lichens

brown trout

water boatman

The table below shows some of the environmental factors in each of six different habitats, A to F.

Environmental Factor	Habitat A	B	C	D	E	F
Oxygen concentration	18%	20%	0.1%	20%	1%	0.8%
Usual temperature range (°C)	− 20 to 5	− 5 to 40	always around 37	always around 35	4 to 14	always around 4
Pattern of light and dark	light and dark periods of varying length	light and dark periods of varying length	always dark	12 hours light, then 12 hours dark	light and dark periods of varying length	always dark
Air conditions	usually windy	often hot winds	N.A.	very humid	N.A.	N.A.
Water conditions	usually frozen	very little	churned up	rains every day	slow moving	still
Concentration of salts in solution	N.A.	N.A.	high	N.A.	very low	high
Altitude range (metres above or below sea level)	+ 3000	0 to + 100	N.A.	0 to + 100	+ 100 to − 5	− 1000

KEY: N.A. not applicable, i.e. the factor is not a part of that habitat's environment
+ above sea level
− below sea level

b (i) In which habitat is there least oxygen?
(ii) In which habitat does the highest temperature occur?
(iii) In which habitat does the lowest temperature occur?
(iv) In which habitat would living organisms be most likely to dry up?
(v) In which habitats could green plants which need light *not* live?

c Which habitat could be
(i) a tropical rain forest?
(ii) an ocean floor?
(iii) an Alpine mountain top?
(iv) a large river?
(v) a desert?
(vi) the inside of the small intestine?

d Below are brief descriptions of three organisms. In which of the habitats A to F would each be most likely to live?

ORGANISM X: Breathes air; can survive in temperatures between − 10°C and 40°C; is well protected against loss of water.

ORGANISM Y: Lives in water; requires light for growth.

ORGANISM Z: Absorbs oxygen from water; requires a temperature above 30°C; absorbs food from surrounding solution.

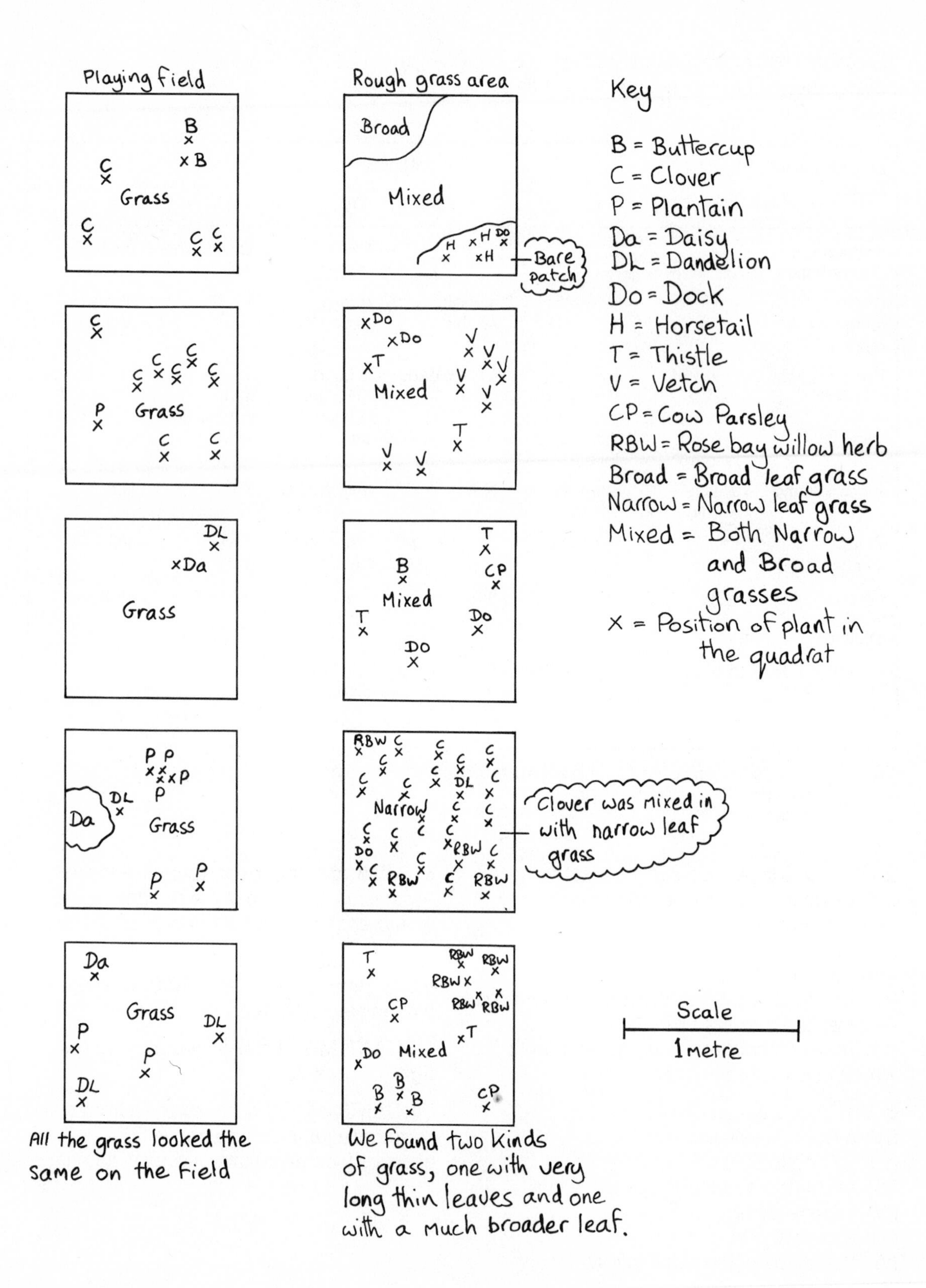

Playing field
Rough grass area
Key
B = Buttercup
C = Clover
P = Plantain
Da = Daisy
DL = Dandelion
Do = Dock
H = Horsetail
T = Thistle
V = Vetch
CP = Cow Parsley
RBW = Rose bay willow herb
Broad = Broad leaf grass
Narrow = Narrow leaf grass
Mixed = Both Narrow and Broad grasses
x = Position of plant in the quadrat
Grass
Broad
Mixed
Narrow
Bare patch
Clover was mixed in with narrow leaf grass
Scale
1 metre
All the grass looked the same on the field
We found two kinds of grass, one with very long thin leaves and one with a much broader leaf.

2 On the playing field

From the laboratory window two students could see the school playing field and an area of rough grass. The only difference seemed to be in the length of the grass. The students wanted to find out if there were any other differences. They used a metre quadrat, a wooden frame measuring 1 m × 1 m. They threw it five times on the playing field and then five times on the rough grass area. Each time they recorded the plants inside the quadrat. The diagram shows a page from Manjit's note book in which she showed the results.

a Which plants were found only on the playing field?

b Which plants were found in both areas?

c Draw a table to show the number of different types of plant found in each of the five quadrats on the playing field and in the rough grass area. Assume the grass on the playing field to be one type, and the narrow and broad leaf grasses to be two different types.

d Now work out the average number of different types of plant per metre squared on the playing field and in the rough grass area.

e What would you conclude from these results?

f Suggest two different reasons to explain the differences between the playing field and the rough grass.

3 London and Delhi

Information about living organisms can help us predict whether the organisms would be able to live in a particular habitat or region. The following table shows the mean monthly temperatures for London and Delhi.

Month	Mean monthly temperature (°C)	
	London	Delhi
January	4	14
February	4	17
March	6	23
April	9	30
May	13	33
June	16	33
July	18	30
August	18	29
September	15	29
October	11	26
November	6	20
December	5	16

a Draw a graph of the mean monthly temperature with both sets of figures on the same axes. Use the horizontal axis for the months, and the vertical axis for the temperatures.

b Describe how the mean temperature in each place varies during the year.

This table shows the temperature range in which three types of insect can develop.

Insect	Temperature range allowing development
Firebrat	25 °C to 48 °C
Housefly	15 °C to 35 °C
Springtail	7 °C to 27 °C

c For each insect, state whether it would be more likely to live in London or in Delhi, or whether it could live in both.

4 On the moors

Bracken and bilberry are two common plants to be found on the moors in the Pennines. Two botanists wanted to find out if the pH of the soil affected the distribution of these two plants. They made random samples at several places. At each place they measured the pH and noted whether bracken or bilberry was present. They could then work out how many times each type of plant was found at a particular pH. They worked out a percentage frequency. The following table shows their results.

pH	Percentage frequencies Bracken	Bilberry
4.8	0	100
4.9	0	100
5.0	0	100
5.1	20	80
5.2	46	62
5.3	47	65
5.4	57	43
5.5	61	64
5.6	53	69
5.7	31	75
5.8	71	14
5.9	44	33
6.0	71	14
6.1	67	0

a Draw two bar graphs exactly one above the other to show this information, on axes like those shown below.

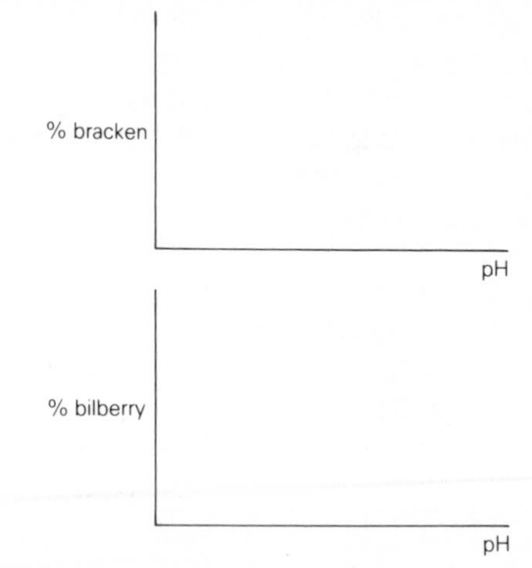

b Which species can grow best in very acid conditions?

c Which species would probably be able to grow better in a soil which has pH 6.5? Explain your answer.

d Explain how it is possible for the two percentage frequencies
(i) to add up to more than 100% (e.g. at pH 5.5);
(ii) to add up to less than 100% (e.g. at pH 5.8).

5 Life in a stream

One of the factors which will affect animals which live in streams is how fast the water is flowing. In an experiment two ecologists found out
(i) the maximum current speed in centimetres per second at which a particular animal is found in a stream;
(ii) the maximum current speed the animal could move against in laboratory tests;
(iii) the current speed at which the animal was washed away in laboratory tests.

The table opposite shows some of the results:

a How many of the species in the list could be found in a stream
(i) with current speed 25 cm/s?
(ii) with current speed 100 cm/s?

b Which animal was not washed away in the laboratory test? Explain how you know.

c Which of the two laboratory tests gives the best guide to the maximum current speed in which the animals could be found in a stream? Explain your answer.

Species	Max. current speed in stream (cm/s)	Max. current speed animal moved against (cm/s)	Current speed which washed animal away (cm/s)
Dragonfly nymph	10	54	77
Flatworm 1 (*Polycelis felina*)	10	44	99
Flatworm 2 (*Planaria alpina*)	14	140	143
Leech	10	37	240
Wandering snail	14	117	202
Freshwater limpet	24	109	240
Freshwater shrimp	40	44	99
Blackfly larva	114	117	240
Stonefly larva	125	100	200
Mayfly larva	222	more than 240	more than 240

6 In the woods

The following table shows the amount of light which reaches the forest floor in a beech forest and in a pine forest during a year. The amount of light is shown in the table as a percentage of the light above the trees. 100% means that it is just as bright on the forest floor as it is above the forest.

Month	% light on the forest floor Beech	Pine
January	85	4
February	90	4
March	90	4
April	80	4
May	10	3.5
June	2.75	3
July	2.75	2.75
August	2	2.75
September	2.25	3
October	3.5	3.5
November	20	4
December	85	4

a Plot these results on a graph, using the same axes, and clearly label the beech forest and pine forest light curves.

b Explain the pattern for
(i) the pine forest floor;
(ii) the beech forest floor.

c What effect would each of these types of trees have on the vegetation growing below them?

d Which forest would have the greater number of different types of plants growing in the soil below the canopy? Explain your answer.

7 Trees and la Montagne Noire

Not far from the famous walled city of Carcassonne in the south of France is the small village of Villegailhenc. Twenty-four kilometres due north is the town of Mazamet and between them is la Montagne Noire (the Black Mountain) which rises to 1210 metres high. At 10.8 kilometres north of Villegailhenc the soil changes from a limestone soil to a sandy soil.

Study the diagram, which shows the profile of the mountain, the trees and the average annual rainfall and temperature.

Diagram showing the zonation of trees over la Montagne Noire in relation to altitude, rainfall, temperature and soil type

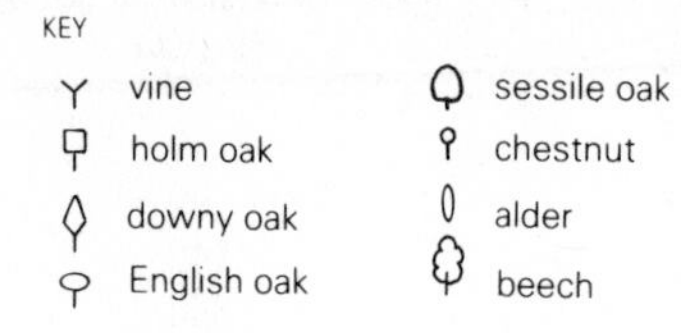

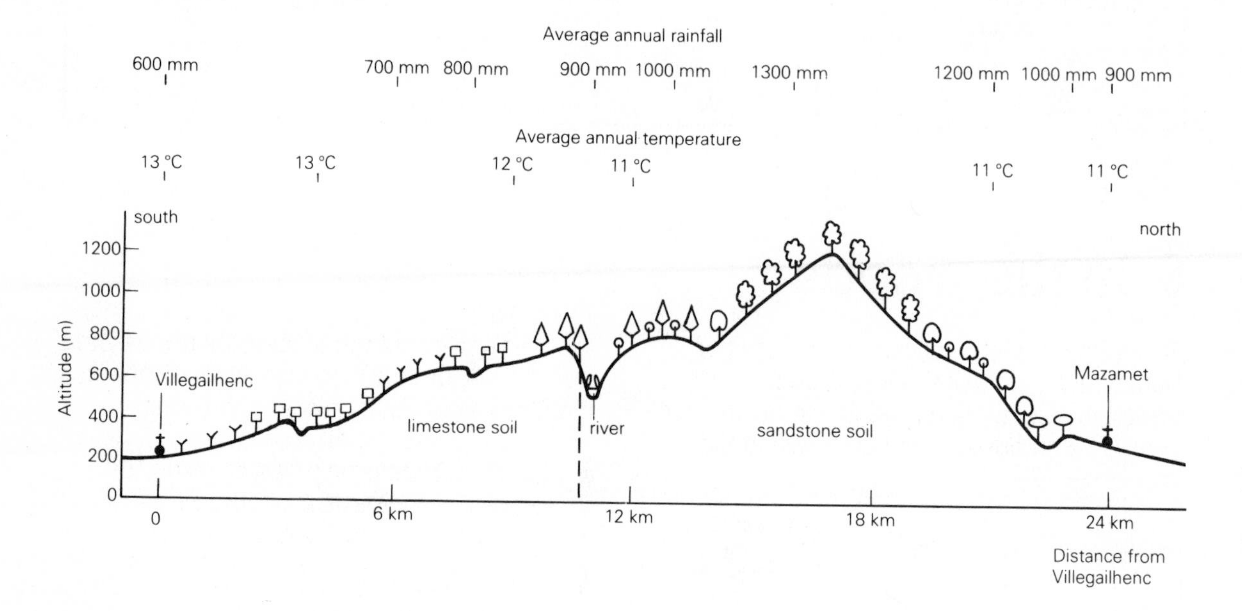

a Which type of tree is found on the summit of la Montagne Noire?

b Which type of tree is always growing with another to form a mixed woodland?

c Which type of tree is able to grow both on a limestone soil and on a sandstone soil?

d What do you notice about the habitat of the alder?

e (i) Why do you think the vines are important to the village of Villegailhenc?
(ii) Suggest why vines are found near Villegailhenc, but not near Mazamet.

f Using the evidence from the diagram, give as complete a description as possible of the conditions under which each of the following grows:
(i) the beech;
(ii) the holm oak.

g What evidence, if any, is there that each of the following factors affects the distribution of trees?
(i) soil type;
(ii) altitude;
(iii) rainfall;
(iv) temperature.

UNIT 2

The variety of living organisms

There are very large numbers of different types of living organisms, many of which look very similar at first glance. Biologists classify them on the basis of careful observation of the similarities and differences. The questions in this unit are about classification and identification of organisms, and possible reasons for the variety.

8 Single-celled animals

The diagram shows ten single-celled organisms. They belong to the group called the Protozoa. They are not drawn to the same scale.

a Look carefully at the similarities and differences between the ten Protozoa. How many subsets can you see?

Call each of your subsets A, B, C and so on, and for each subset write down the numbers of the Protozoa that you have put in the subset.

b For each subset, say what is or are the main feature(s) that made you put the Protozoa in that subset.

9 Vertebrate classification

The vertebrates are classified into five main groups, called *classes*. Each class is divided into smaller groups, called *orders*.

Both of the animals shown opposite belong to the same class, but they belong to different orders.

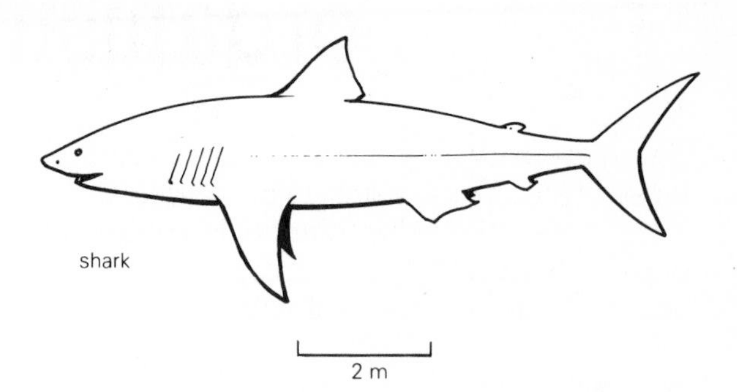

a Which class do these animals belong to?

b List all the features which these animals have in common.

c List all the features which you can see that would place these two animals in different orders.

d Using the scales, calculate the actual lengths of these two animals, from the snout to the notch in the tail.

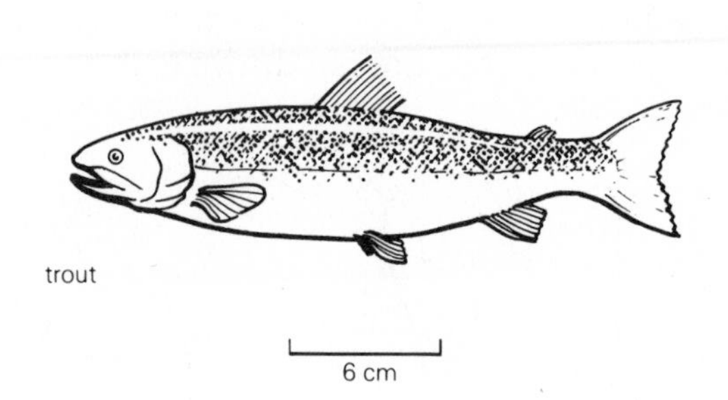

10 Stonefly nymphs

Stoneflies are insects which live near streams. The eggs are laid in the streams and hatch into larvae, called nymphs. These nymphs crawl about on the stones in the stream, often on their underside so that they are not swept away by the current. When a nymph is fully grown it crawls out of the water and changes into an adult.

The drawings show the nymphs of nine species of stonefly. Species X is the same as one of the species A to I, although it has been drawn a little larger.

a Which of the species A to I is the same as species X?

b Which features of species X did you find most helpful in matching it with the similar drawing?

c Work out the magnification of the drawing of species X.

d Draw a table to show the actual *body* length of each species, A to I.

e Use your table to say which nymph is the smallest and which is the largest.

Study specimens F and G.

f List the similarities between nymph F and nymph G.

g In a table, list the differences between nymph F and nymph G.

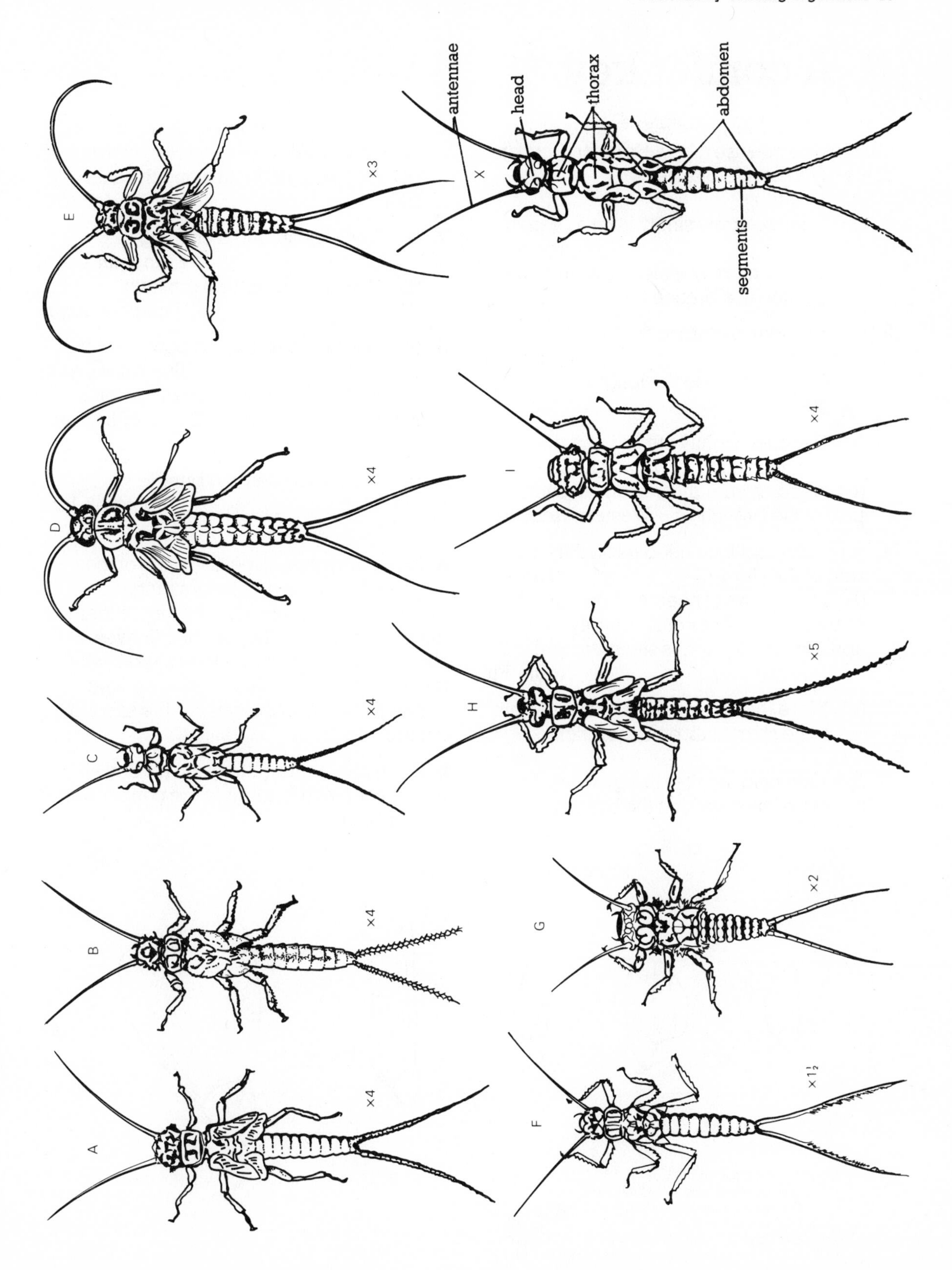
antennae
head
thorax
abdomen
segments
X
A
×4
B
×4
C
×4
D
×4
E
×3
F
$\times 1\frac{1}{2}$
G
×2
H
×5
I
×4

11 A conifer key

In order to identify species of plants or animals we can use a key. Here is a key for some of the coniferous trees.

1. (a) Needles grow singly from a branch (go to 2)
   (b) Two or more needles grow from the same point on a branch (go to 5)
2. (a) No clear pattern to the length of the needles (go to 3)
   (b) Clear pattern to the length of the needles (go to 4)
3. (a) Needles slightly curved and point along the branch **Mountain Hemlock**
   (b) Needles straight and fan out each side of the branch **Western Hemlock**
4. (a) Short and long needles lie flat each side of the stem **Giant Fir**
   (b) Many rows of needles along the branch. Outer ones curl strongly upwards, middle ones stand vertical **Noble Fir**
5. (a) More than ten needles growing from the same short stalk on the branch (go to 6)
   (b) Less than ten needles growing from the same short stalk on the branch (go to 9)
6. (a) Needles all grow out from the stalk at the same level (go to 7)
   (b) Needles grow out along the short stalk (go to 8)
7. (a) Needles around 2 cm long **European Larch**
   (b) Needles around 3.5 cm long **Japanese Larch**
8. (a) Needles curved and point forwards along the stalk **Blue Atlas Cedar**
   (b) Needles straight, sharply pointed and stiff **Cedar of Lebanon**
9. (a) Two needles in a bundle **Scots Pine**
   (b) Three needles in a bundle **Monterey Pine**

a In the drawings opposite, parts of the five trees are shown. The drawings are more or less full size. Use the key to identify the five trees from the needles (leaves) shown in the drawings. Draw a table with the letters A to E in one column and the name of the correct conifer in the second column.

b If you found this quite easy, try devising a key of your own to identify the conifer seeds shown below.

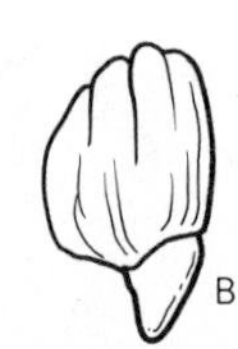

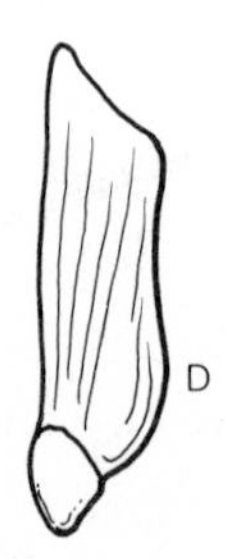

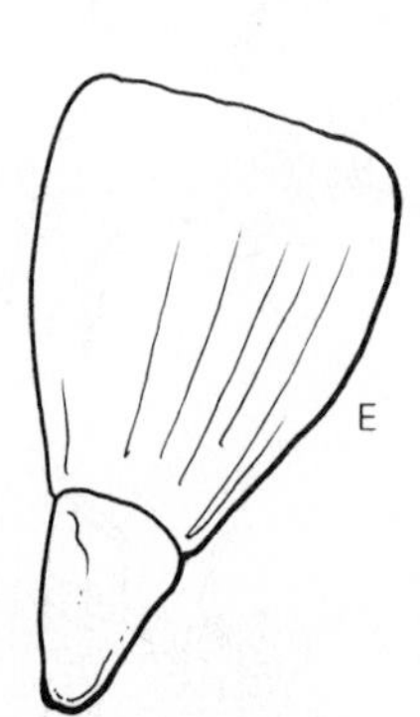

Actual size

A = European larch
B = Japanese larch
C = Sitka spruce
D = Scots pine
E = Giant fir

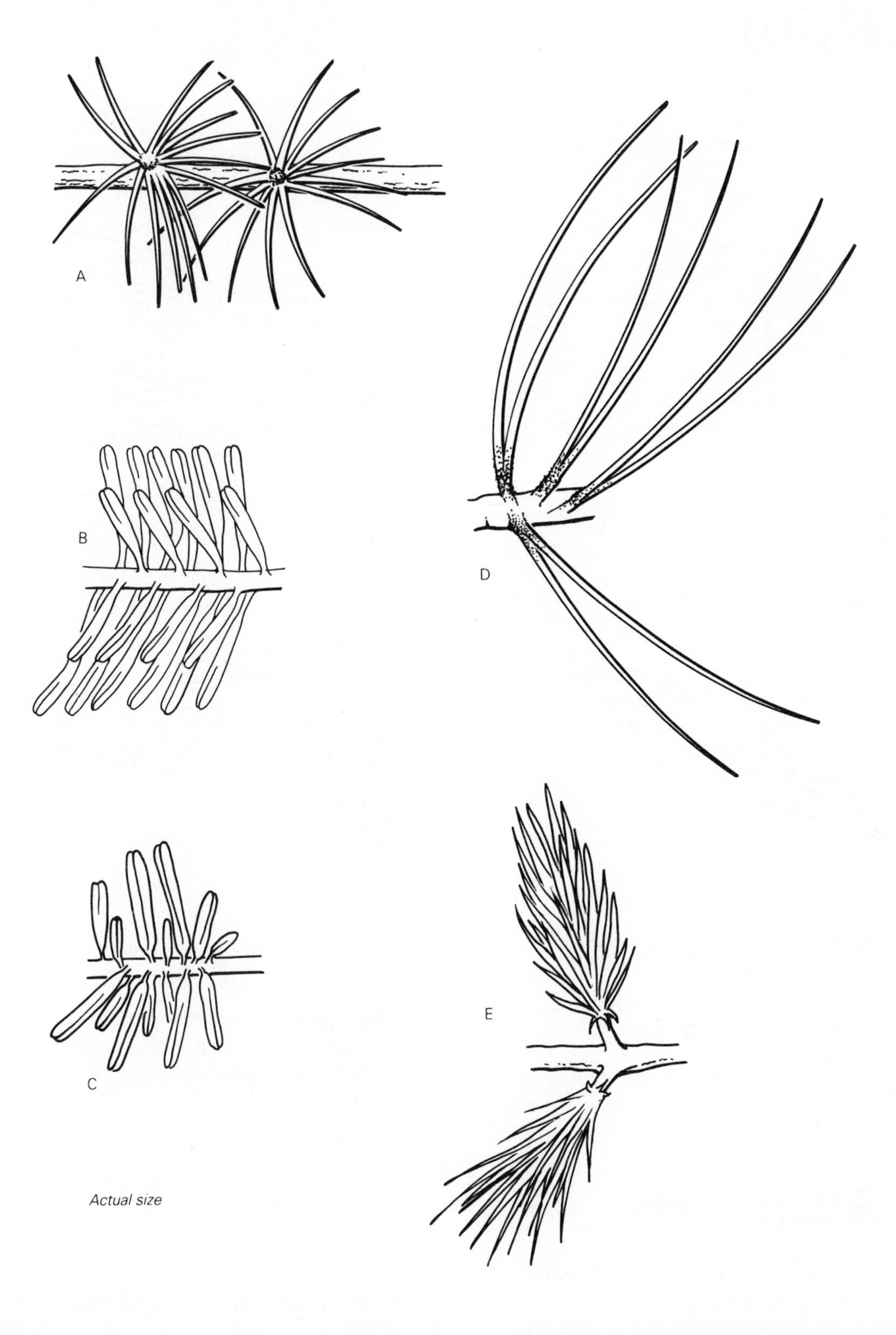
A
B
C
D
E
Actual size

12 Flying

The drawings show structures which enable an animal to use air for moving about. They are not drawn to the same scale.

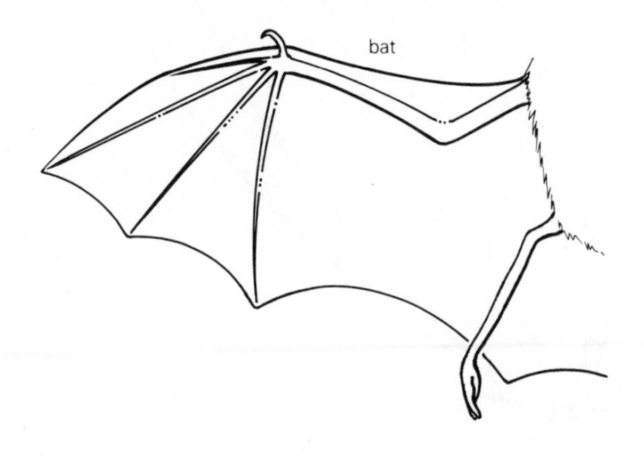

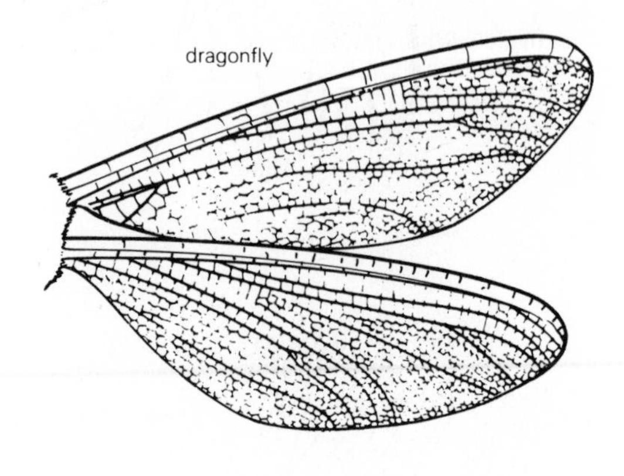

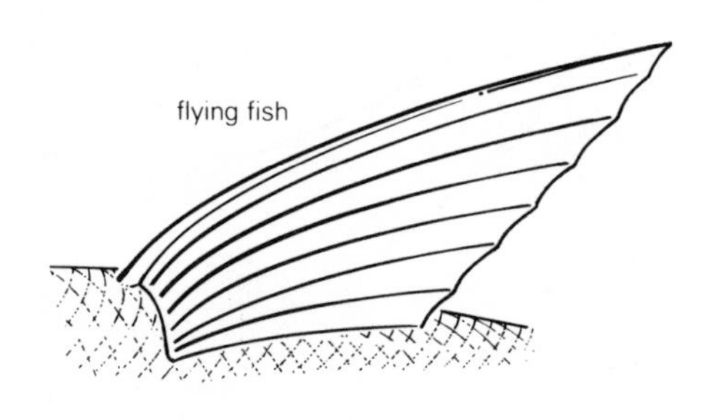

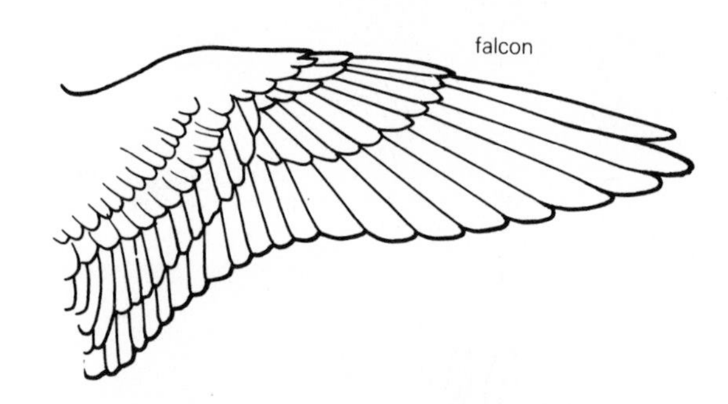

a (i) Describe two features all the structures have in common.
(ii) Suggest how each feature helps in moving the animal through the air.

b Apart from size, what are the main differences between:

(i) the dragonfly and flying fish structures?
(ii) the bat and the falcon structures?

c The flying fish uses its structures for gliding. Suggest how the structures differ from ones used for flapping flight.

13 Adaptation

The drawings below show ten different organisms. Each type of organism has certain special features, called adaptations, which enable it to live in a particular set of conditions.

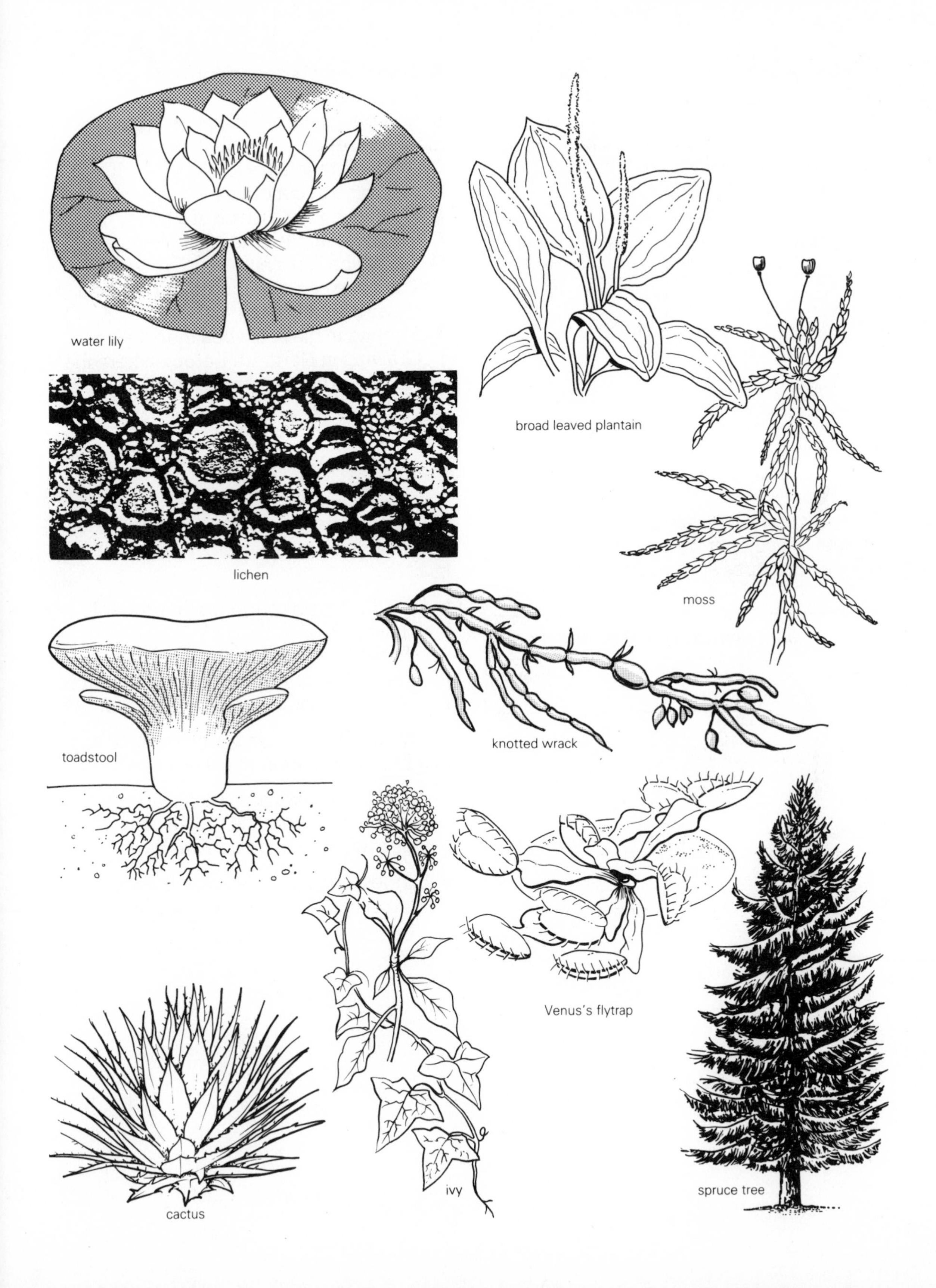
water lily
broad leaved plantain
lichen
moss
toadstool
knotted wrack
Venus's flytrap
cactus
ivy
spruce tree

Draw a table like the one below and match each organism to the letter of the statement which shows a set of conditions to which the organism is adapted. In each case suggest one special feature which adapts the organism to these conditions.

Organism	Set of conditions	Special adaptation for conditions

A Completely bare rock on a mountain-side in Scotland.
B Completely water-logged soil on the moors in Ireland.
C A dead branch in a forest.
D Semi-desert conditions.
E Floating on the surface of a pond.
F Climbing up the side of a building.
G Soil which lacks nitrates.
H The edge of a public footpath across a playing field.
I Floating in the sea when the tide is in.
J Growing in Northern Europe where the winters bring heavy snowfalls.

14 The naked ape?

The following table shows the average number of hairs per square centimetre of skin on adult apes and human beings.

	Average number of hairs per cm^2		
Adult	**Crown of the head**	**Back**	**Chest**
Apes			
Chimpanzees	185	100	70
Gorillas	410	145	4
Orangutans	158	175	105
Human beings			
Europeans	330	0	3
Mongolians	128	0	0
Negroes	305	0	0

a Describe the main differences between apes and human beings in the distribution of hair.

b (i) Which type of ape would probably look hairiest?
(ii) What other information would you need in order to find out which type of ape really has most hair on the upper part of its body?

c Which of the human types has least hair and is closest to being a 'naked ape'?

d A European man has about 500 cm^2 covered with hair on the crown of his head. About how many hairs does he have?

e Suggest two possible advantages of the hair density being so high on the head.

UNIT 3 Ecosystems

The plants and animals in a particular habitat form a community which uses and re-uses the resources of the habitat. The organisms interact with each other and with the physical environment. The community of organisms and the environment together make up an ecosystem, such as a pond or a wood. The questions in this unit are about the relationships within such ecosystems.

15 Wytham Wood

Wytham Wood is a wood in the south of England, near Oxford, which has been studied in great detail by biologists. Study this simplified food web for Wytham Wood.

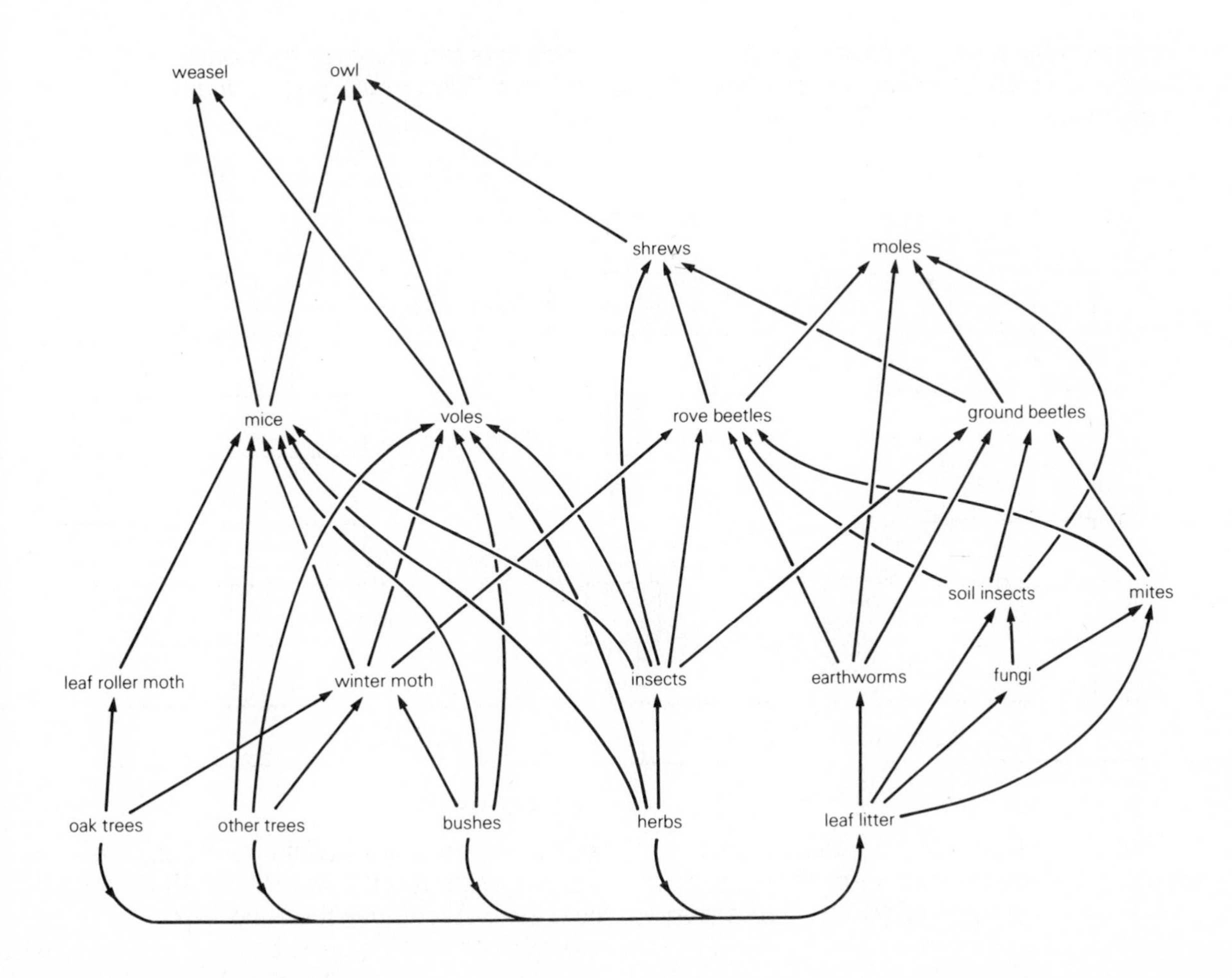

Use the food web to answer the following questions.

a What do owls feed on?

b What do mites feed on?

c Which moth feeds only on one kind of tree?

d Name two animals which are herbivores.

e Name two animals which are omnivores.

f Name two animals which are carnivores.

g Copy and complete these two food chains
(i) herbs → ? → rove beetles → ? → ?.
(ii) ? → ? → mites → ? → ?.

h How many primary consumers are shown in the food web?

i Name four animals which feed at more than one trophic level, and say what the trophic levels are in each case.

j In some years there are very large numbers of leaf roller moths and winter moths. What effect would you expect this to have on
(i) the plants?
(ii) other animals?

16 Pyramids of numbers

The diagrams show the pattern of the numbers of plants and animals in a food chain in the form of a pyramid. The producers are shown as the bottom of each pyramid. The pyramids are not drawn to scale.

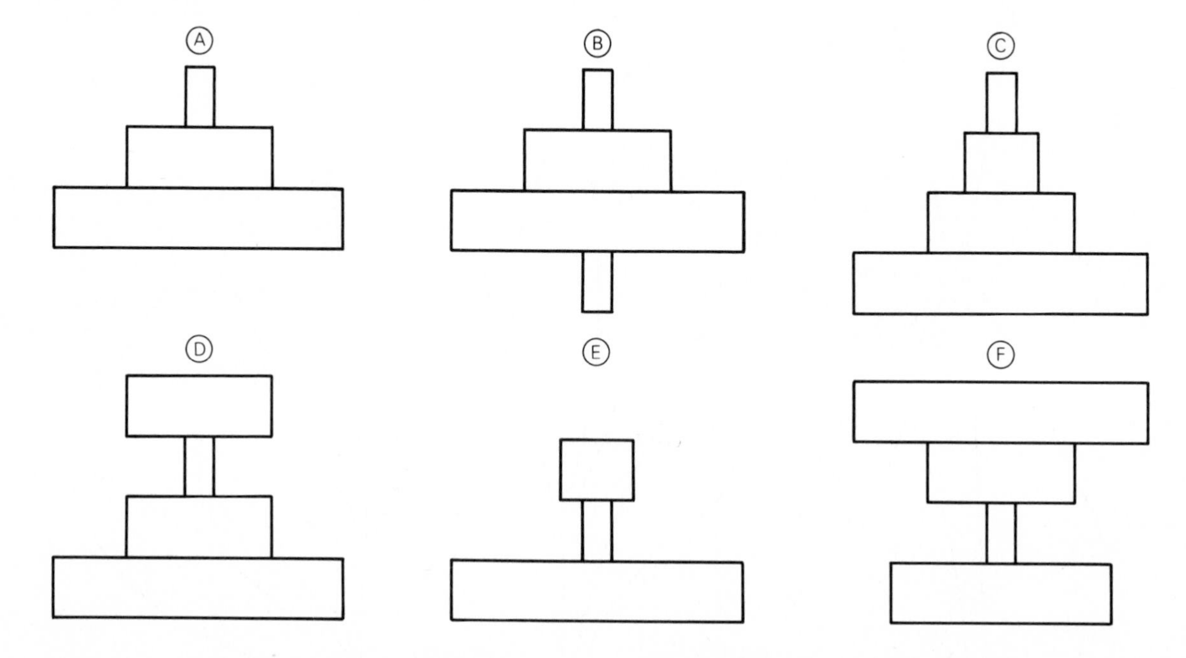

For each of the following descriptions, work out a food chain or web. Then match it with one of the pyramids of numbers shown above.

a The Indian water buffalo was eating the lush vegetation at the waterhole. All the time its tail kept flicking at the hundreds of mosquitoes feeding on its back.

b Out on the African grassland a vast herd of wildebeest was grazing. A female lion crept up to the edge of the herd and, as the herd started to stampede, pulled down and killed a young wildebeest.

c The blackbird, after gorging itself on ripe berries, was preening itself and trying to remove some of the large numbers of lice which were feeding on its skin. What was interesting was that each of the lice had a large number of single-celled protozoa living in its gut.

d The cod was swimming fast into the shoal of herring to feed. The herring had been feeding on the large number of tiny crustaceans, the zooplankton. Zooplankton feed on the enormous numbers of floating algae in the water, the phytoplankton.

e The gardener was spraying his rose bush to try and get rid of the enormous number of greenfly covering the young stems. He was being helped by a few ladybirds. In the garden he saw a blue tit swallow a ladybird.

f The hedgehog was rooting through the leaf litter, eating earthworms, beetles and spiders. Every so often its rear leg would scratch vigorously at its side because of the large number of fleas on its body.

17 Pyramids of biomass and energy

In question 16 we looked at pyramids of numbers. Some of the diagrams are not really pyramid shaped because the pyramid is based only on the numbers of organisms. For example, there may be two or three hundred caterpillars feeding on one cabbage plant. A pyramid of numbers for this situation is shown opposite. However, if we measure the *mass* of these two organisms we get a true pyramid pattern – a pyramid of biomass.

The following table shows an analysis of a warm water ecosystem in Florida.

Trophic level	Examples of organisms	Dry biomass (g/m^2)	Energy (kJ/m^2)
Tertiary Consumers	fish, alligators	1.5	35
Secondary Consumers	birds, frogs, insects, fish	11	220
Primary Consumers	molluscs, crustaceans, fish, turtles	37	630
Producers	algae, aquatic plants	809	14 600

a Why is it better to use *dry* biomass as a measure rather than fresh biomass?

b For this ecosystem in Florida make accurate diagrams on graph paper to show:
(i) a pyramid of biomass;
(ii) a pyramid of energy.
Your diagram shuld be drawn to a suitable scale, e.g. $1\ cm^2 = 25\ g/m^2$.

c How do you think ecologists are able to estimate the energy in a given trophic level?

18 Ecosystems and energy

From question 17 it can clearly be seen that the amount of energy in each trophic level rapidly decreases from the producers to the tertiary consumers. Below is some data on the 'energy budget' for a forest ecosystem. This shows how much energy each square metre of forest received in one year, and what happened to the energy used by the trees.

Total energy received from sunlight
4 000 000 kJ/m^2/yr

Total energy gained by photosynthesis
50 000 kJ/m^2/yr

Total energy lost through respiration
39 000 kJ/m^2/yr

This information is shown in diagrammatic form, right.

a Calculate the percentage of the sunlight energy received which is fixed by the trees in photosynthesis.

b Calculate how much energy is used for new growth by the trees.

c Calculate the percentage of the energy fixed in photosynthesis which is used for new growth.

d Suggest what happens to the energy from sunlight which is received by the forest but is not used in photosynthesis.

This picture of only a small amount of the energy taken in by an organism being used for growth is repeated at each trophic level. For example, consider this information for a weasel feeding on voles.

Energy available to the weasel	22 000 kJ
Energy in voles but not eaten	1065 kJ
Energy in weasel's waste	53 kJ
Energy released in respiration	20 400 kJ
Energy used for growth of weasel	482 kJ

e How much of the energy available to the weasel is actually taken in as food?

f What percentage of the energy actually taken in by the weasel is used for new growth?

g What percentage of the energy in the food eaten by the weasel is lost in its waste?

h Explain why some of the energy available in the voles would not be eaten by the weasel.

i Suggest four ways in which the weasel would use the energy from respiration.

j Draw a pie chart or some other kind of diagram to show the energy budget of a weasel.

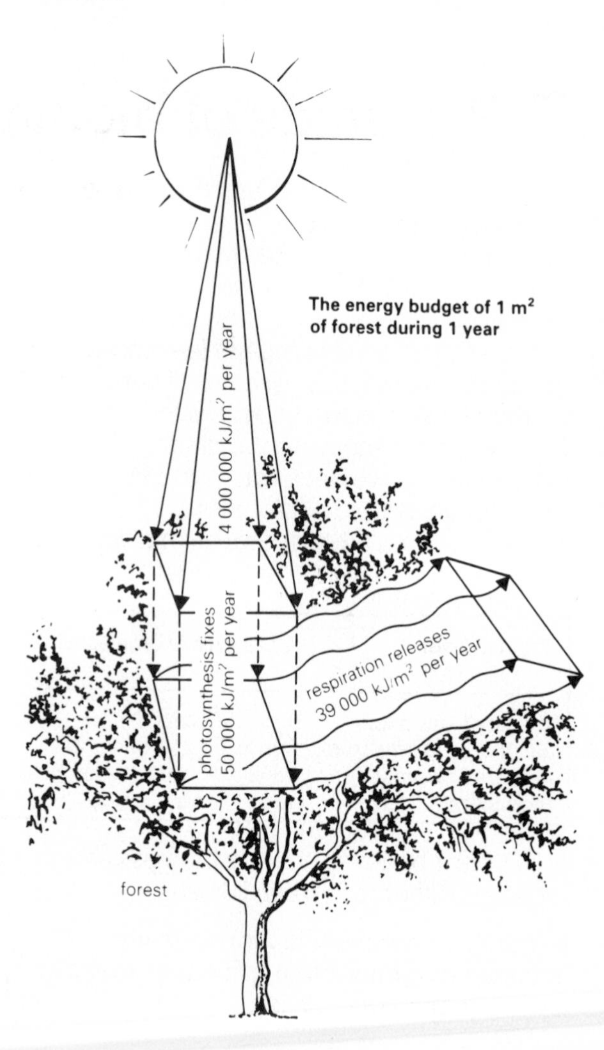

The energy budget of 1 m^2 of forest during 1 year

19 In the desert

A community is all the organisms living together in a particular habitat. The photograph shows a desert community in Africa.

With no plants growing in the area, the community relies for energy on seeds and dry plant fragments blown across the desert. Ants, bristletails, beetle larvae and termites all feed off the seeds and plant material. The adult dung beetles feed on dung. The ants are eaten by antlion larvae which lie nearly buried in the sand except for their jaws. Spiders feed on the ants and bristletails. Worm snakes feed on the termites. Scorpions eat beetle larvae and spiders whilst the camel spider is able to devour scorpions as well as beetle larvae and dung beetles. Lizards eat spiders, dung beetles and camel spiders.

a In this desert community, name
(i) three primary consumers;
(ii) three secondary consumers;
(iii) two tertiary consumers.

b Draw a food web to show all the feeding relationships.

c From this food web, draw separate food chains
(i) involving two animals only;
(ii) involving three animals only;
(iii) involving four animals only;
(iv) involving five animals.

d Which will be the most common of the animals in this community, and which will be the least common? Explain your answer.

e Explain how this community ultimately depends on the sun's energy for its survival.

Namib desert

20 An oak wood

The following table shows the change in biomass in an oak wood over one year. The measurements were taken from an area of 40 m^2.

Region	Biomass when first measured (kg/m^2)	Biomass 1 year later (kg/m^2)
Oak tree canopy	954	977.6
Shrub layer	72.4	80.4
Herb layer	2.8	5.2
Fallen leaves	0	14
Roots	221.6	230.8

a Draw a table to show the percentage increase in biomass during one year for each region of the wood.

b From your percentage table say
(i) which region has the highest net production;
(ii) which region has the lowest net production.

c Do you think the figures for increase in biomass after one year are likely to be accurate? If not, are they more likely to be overestimates or underestimates? Explain your reasons.

21 Discovering feeding relationships

In the previous questions a number of statements were made about the diet of various animals.

a Suggest two different methods which may be used by ecologists to find out what an animal feeds on.

One method used by an ecologist studying a salt marsh community was to use a radioactive isotope, phosphorus-32 (^{32}P). He injected the isotope into the living grass in one area and the dead plant material (detritus) in another area. He then measured the radioactivity in the animals living in the community. Below are two graphs showing some of his results. Graph A shows the results for an insect and Graph B shows the results for a snail.

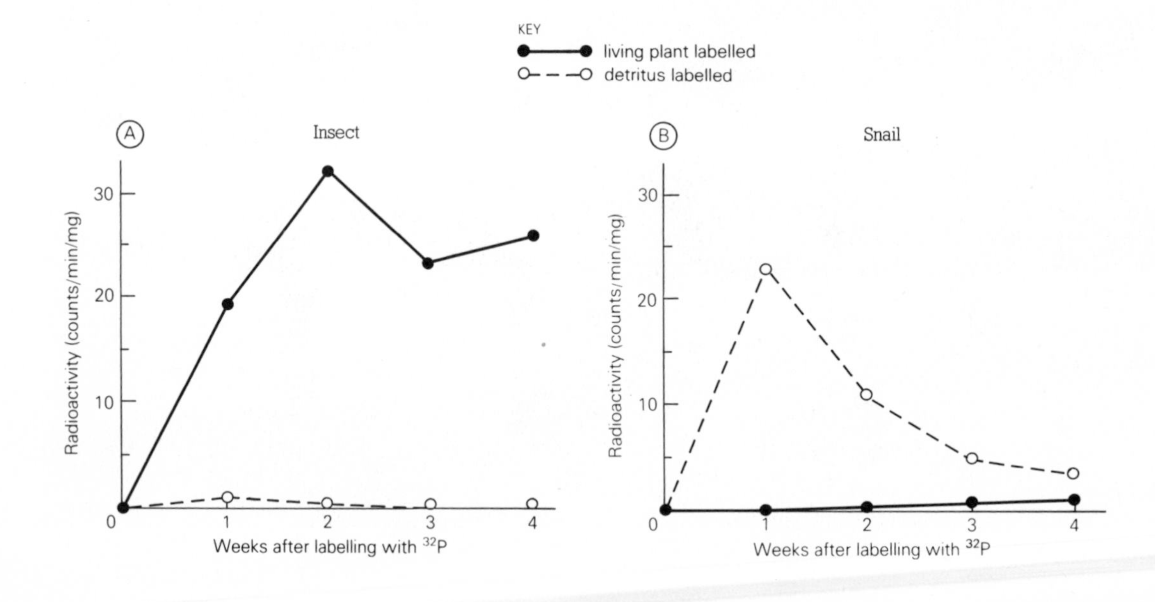

b What do you conclude is the food of the insect? Explain your answer.

c What do you conclude is the food of the snail? Explain your answer.

d The ecologist found that the radioactivity took longer to appear in spiders than it did in either insects or snails.
(i) Explain why it took longer for the radioactivity to appear in the spiders.
(ii) What do you conclude about the diet of the spider?

22 Predator–prey relationships

In a laboratory investigation, plant mites were provided with oranges for food. Some carnivorous mites were also put in with them. Every six days the number of plant mites and carnivorous mites were estimated. The results are shown in the table below.

Day	Number of plant mites	Number of carnivorous mites
1	50	2
6	250	2
12	800	8
18	1400	30
24	700	40
30	300	18
36	250	16
42	100	10
48	200	8
54	200	6
60	300	2
66	500	2
72	600	8
78	1600	6
84	1800	20
90	2000	18
96	1800	28
102	1400	22
108	750	40
114	500	10
120	750	4

a Draw graphs to show the changes in the numbers of plant mites and carnivorous mites.

b What do you notice about the population peaks of the plant mites and the carnivorous mites?

c (i) What do you think caused the decline in the plant mite population between day 19 and day 42?
(ii) What evidence is there to support your idea?

d What enabled the carnivorous mite population to increase from day 66 through to day 108?

e What caused the decline in the carnivorous mite population from day 108 through to day 120?

f What conditions are necessary for the plant mite population to increase?

23 Populations

A number of ring-necked pheasants were introduced onto an island. There were no predators of pheasants on the island. Study the graph, which shows the increase in the pheasant population over six years.

a You will notice that each year the population of pheasants increases between April and September and then there is a decrease from September to the following April.

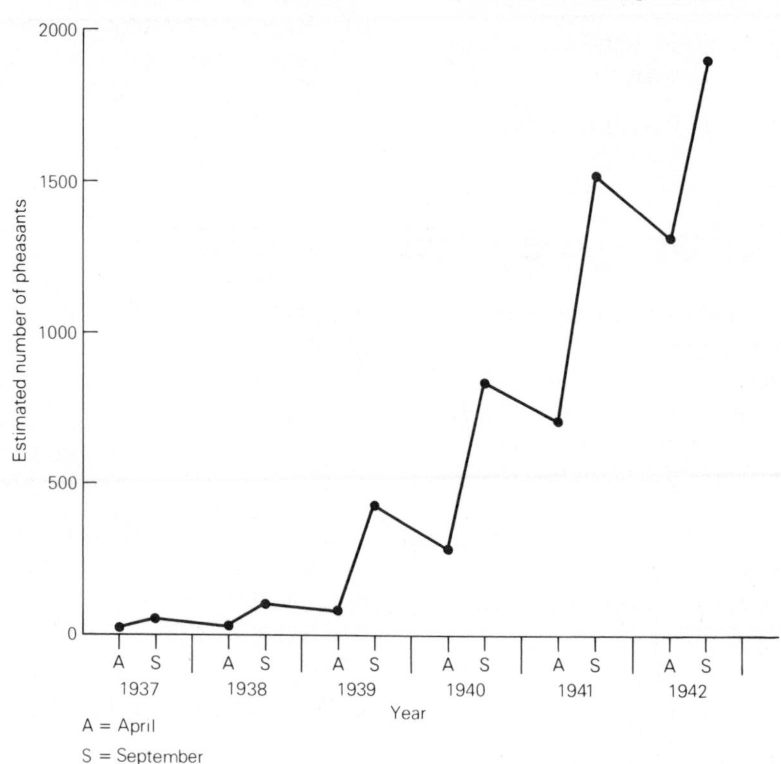

(i) Suggest why the population increases each year.
(ii) Suggest a reason for the decrease each year.

b In which year was there the biggest increase in the pheasant population?

c What evidence is there that the pheasant population growth was beginning to slow down by 1942?

d Suggest two reasons why the pheasant population growth was beginning to slow down.

The following table 'smooths' out the growth curve by taking the number of pheasants half way between the September peak and the following April trough.

Year	Smoothed population growth
1937	8
1938	20
1939	90
1940	354
1941	775
1942	1433

e Plot a graph to show the smoothed population growth of the ring-necked pheasants. Continue the time axis up to 1943.

f Describe the pattern of the population growth and explain why it is a 'curve' rather than a straight line.

g Continue your graph with a dotted line to show what you think happened to the pheasant population in 1943. Explain why you have drawn your extrapolation as you have.

UNIT 4 Natural cycles

The organisms in an ecosystem require a variety of nutrients from the environment. Where man artificially manages an ecosystem and removes organisms, some of these nutrients may have to be replaced. The questions in this unit are about the cycling of natural resources and ways in which man may affect these natural cycles.

24 The water cycle

a Using only the information in the following passage, construct a diagram which summarises this description of the water cycle.

The heat energy from the sun causes water to evaporate from the oceans, lakes and rivers. The water vapour rises and, as it does so, cools and condenses to form clouds. Plants also add to the water vapour in the atmosphere through transpiration. Clouds release water as rain which falls to the ground and is either taken up by plants or runs off by way of rivers back to lakes and the sea.

b Explain why cutting down large areas of forest could have a marked effect on the local water cycle. What effect would it probably have?

c Water is an important limiting factor in many parts of the world, so that farmers must rely on irrigation to supply the water needed for growing crops. However, irrigation can cause very serious problems. The map shows the Indus river basin in Pakistan.

13 000 000 hectares were irrigated *but* poor drainage caused waterlogging of the soil. Mineral salts in the irrigation water

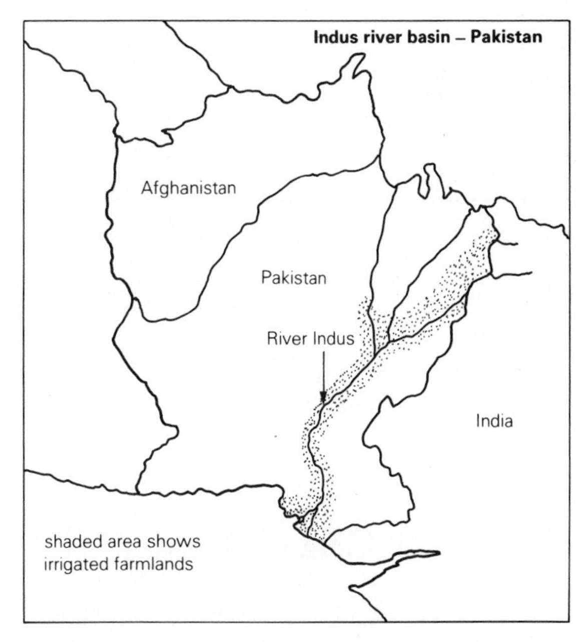

were not washed out of the soil, and became so concentrated that productivity was destroyed or reduced over 6 000 000 hectares. The Pakistan government estimated they were losing one hectare of productive land every 12 minutes.

(i) Why do waterlogging and increased salt concentration in the soil make the land infertile?

(ii) Suggest how the farmers could lower the water table to make the land productive again.

d Engineers, scientists and economists have to provide governments with information to enable people to make sensible decisions about the kinds of schemes which should be started to help manage water. In some areas it may be possible to construct either one large or several small reservoirs.

Consider the information in the table overleaf, which compares two plans. One plan is to build a single large reservoir in an area; the other is to have 34 small reservoirs in the same area.

	One large reservoir plan	Small reservoirs plan
Number of reservoirs	1	34
Drainage area (hectares)	50 500	49 200
Flood storage (hectare metres)	6 400	7 300
Surface area for recreation (hectares)	790	850
Valley bottoms flooded (hectares)	750	650
Valley bottoms protected from flooding (hectares)	1 365	3 270
Total cost (dollars)	6 000 000	1 983 000

Note: A hectare metre is equal to an area of one hectare flooded to a depth of one metre.

(i) Explain all the reasons why having 34 small reservoirs is better than having one large one.
(ii) Suggest one possible problem with having 34 small reservoirs.

25 The carbon cycle

It has been calculated that each year a total of 166 gigatonnes of carbon is cycled throughout the world. (A gigatonne is one thousand million tonnes.) The pie chart opposite shows the proportions of the carbon transferred by different processes.

a Calculate the amount of carbon fixed by photosynthesis.

b Calculate the total amount of carbon returned to the atmosphere through the respiration of living organisms.

c What are the main processes which contribute to the amount of carbon generated by combustion?

d In 1900 the percentage of carbon dioxide in the atmosphere was around 0.029%. In 1950 it was 0.030%. In 1983 it was 0.0325%.
(i) Draw a graph and use it to estimate the percentage of carbon dioxide in the year 2000.
(ii) This trend could result in the 'greenhouse' effect. Write a paragraph which sets out the main ideas behind the greenhouse effect, which could be understood by a 10 or 11-year-old.

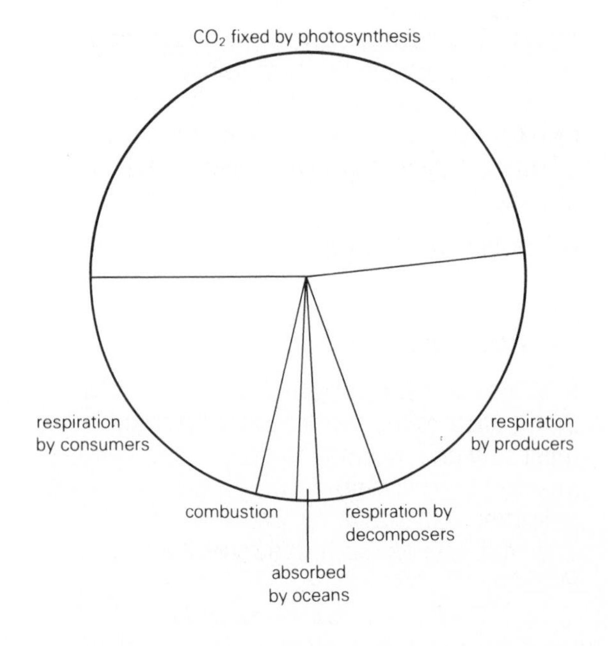

Total global transfer of carbon = 166 gigatonnes

26 The nitrogen cycle in a meadow

Study this diagram of the nitrogen cycle in a meadow.

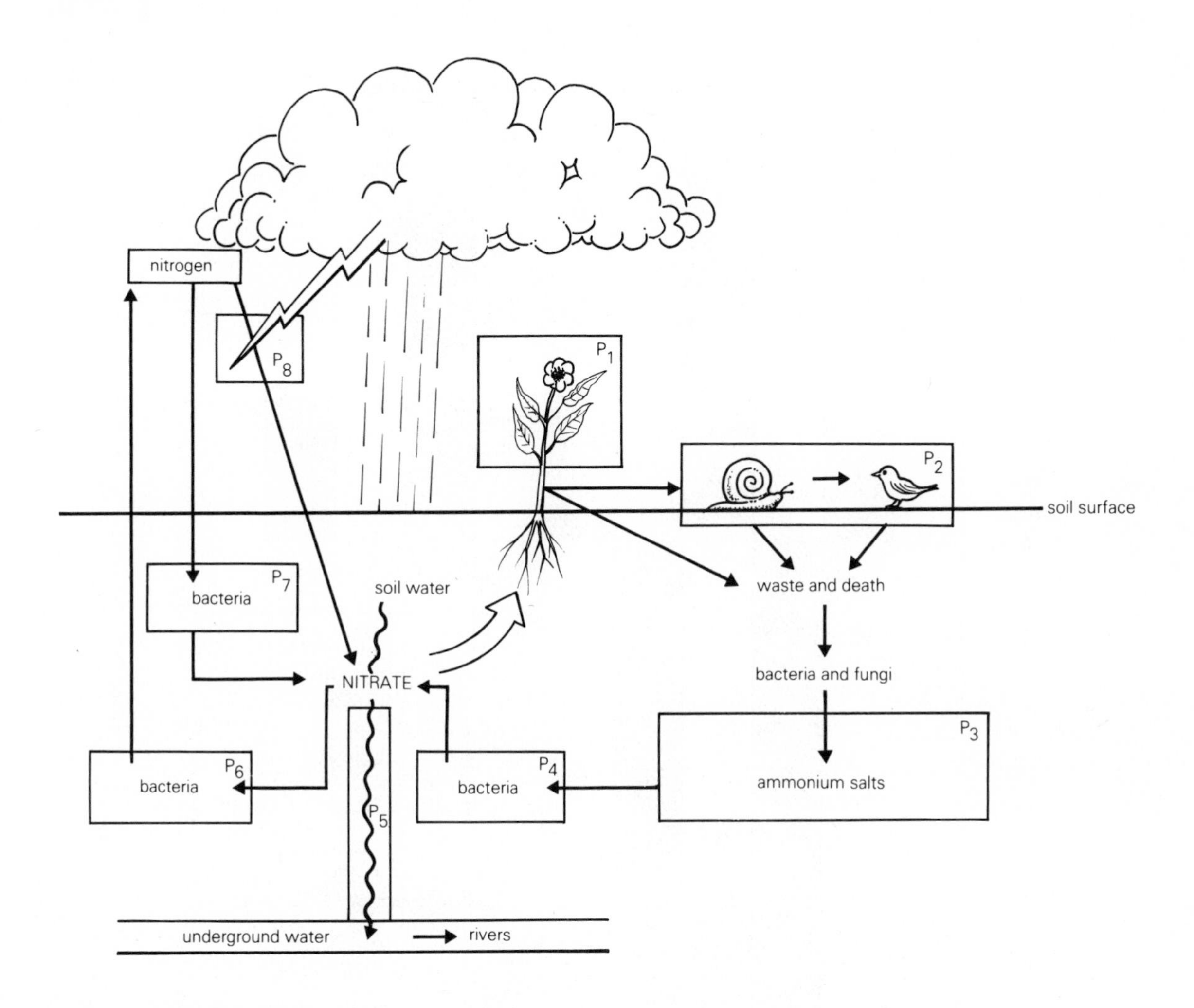

a In how many ways is nitrate added to the soil?

b In how many ways is nitrate removed from the soil?

c The boxes marked P_1 to P_8 represent the processes which go on in the nitrogen cycle. These processes can be described as follows:

feeding leaching denitrification lightning fixation ammonification photosynthesis nitrification nitrogen fixation

Write down the processes in the order shown by the boxes, e.g. P_1 = photosynthesis.

d One important way in which nitrate is added to agricultural land is not shown on this diagram.
(i) What is this addition?
(ii) Why is it necessary?

27 The nitrogen cycle budget

Study the following diagram, which shows the nitrogen cycle budget for land in the USA.

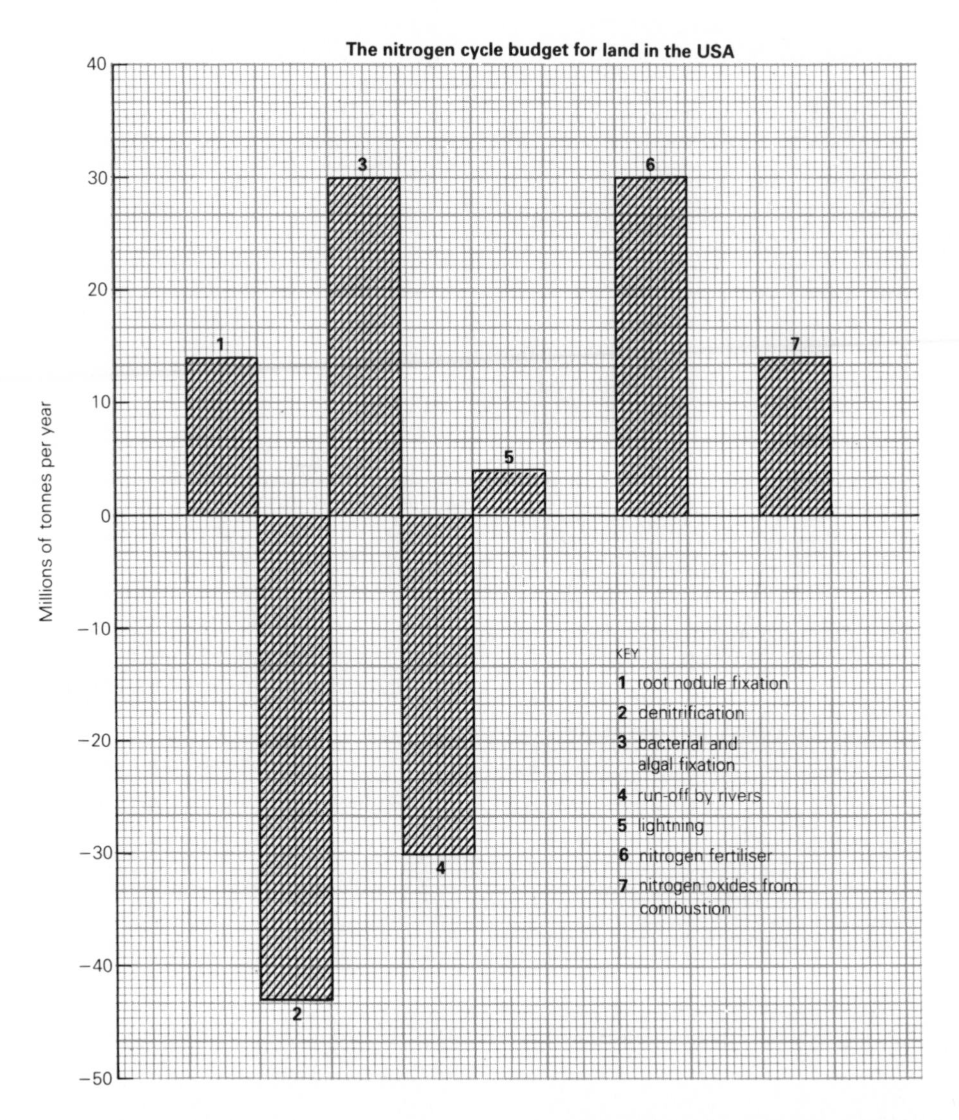

a How much nitrogen is fixed in a year by root nodules?

b Explain why the diagram shows denitrification and run-off *below* the horizontal axis.

c What is the net balance per year in the nitrogen cycle budget

(i) including the influence of human beings?

(ii) excluding the direct influence of human beings?

d Which of the factors 1 to 5 would change the most if human beings stopped affecting the nitrogen cycle budget? Explain your answer.

28 Nitrogen fixation in root nodules

Nitrogen-fixing bacteria enter the root hairs of plants like clover, peas and beans and form swellings on the roots called nodules.

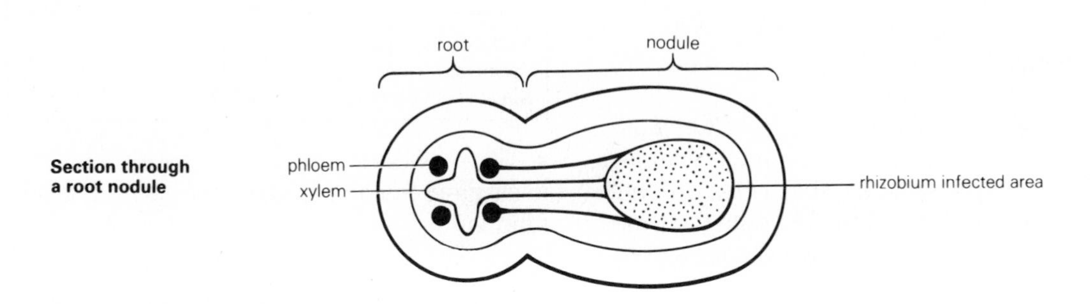

Farmers inoculate red clover seeds with nitrogen-fixing bateria before they sow the seeds. The drawing opposite shows two clover plants grown from seeds planted at the same time. Only one seed had been inoculated.

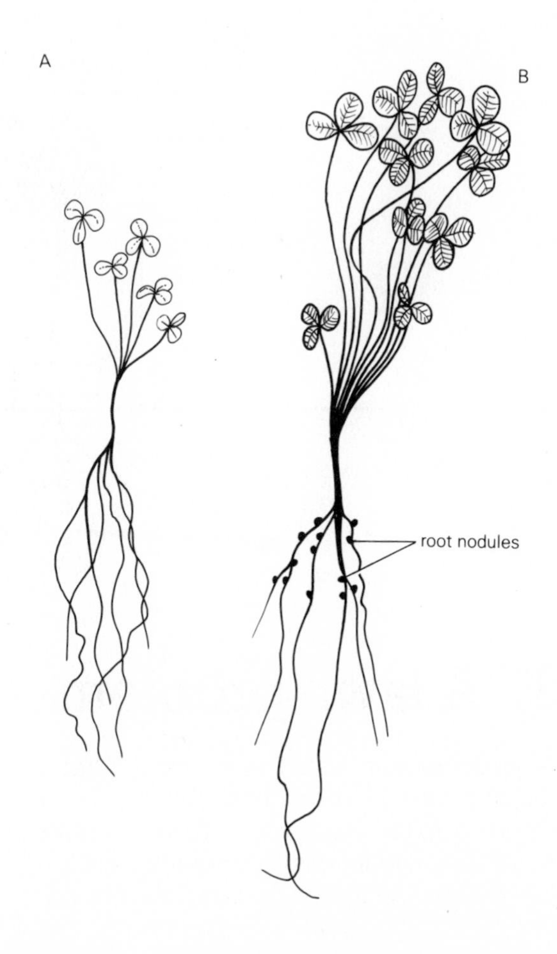

a How many nodules has plant B?

b (i) How many leaves has plant A?
(ii) How many leaves has plant B?

c Which plant came from the seed inoculated with the bacteria?

Symbiosis means that two organisms benefit from living together.

d What evidence is there that the clover plant benefits from the bacteria?

e Suggest one substance the clover plant obtains from the bacteria.

f Suggest one substance the bacteria must obtain from the clover.

g Suggest how substances obtained from the bacteria reach the leaves of the clover.

h Suggest how substances produced by the clover reach the bacteria.

29 Fungal symbiosis

Many plants grow much better when their roots are associated with a particular fungus in the soil. This association between a species of plant and a particular fungus is called a mycorrhiza.

The diagram below shows the apparatus used in an experiment to investigate the association between a fungus and pine seedlings.

Radioactive carbon dioxide ($^{14}CO_2$) was introduced into the pine seedling container. Radioactive phosphate ($^{32}PO_4$) was present in the agar medium. After 12 hours the pine seedlings contained radioactive phosphate, and the fungal threads contained radioactive sugars.

In another experiment with maize growing on soil which had very little phosphate present, the following results were obtained.

	Maize without mycorrhiza	Maize with mycorrhiza
Number of grains per ear	31	354
Mass of 1000 grains	2.4 g	19.8 g

a What evidence does this experiment give to suggest that the maize benefits from the fungus?

b What evidence is there that the fungus obtains benefits from the pine seedlings?

c Explain how radioactive sugars came to be found in the fungal threads.

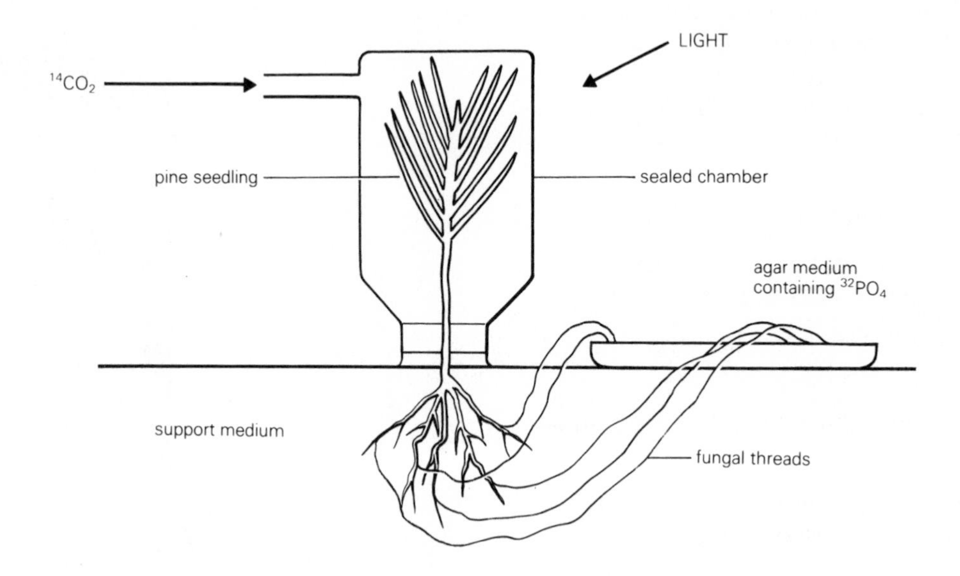

30 A lawn problem

A gardener wanted to make a lawn. After digging over the ground and raking it level she went to the Garden Centre to buy grass seed. The instructions on the packet said that before she sowed the seed she should spread a general fertiliser. The Garden Centre advised that 'Growmore Fertiliser' would be what she needed. The container gave her the following information about the fertiliser.

GROWMORE

COMPOUND FERTILISER 7–7–7

Nitrogen (N): 7%
Total Phosphorus pentoxide (P_2O_5): 7% (P = 3.0)
Soluble in water: 6.5% (P = 2.8)
Insoluble in water: 0.5% (P = 0.2)
Potassium oxide (K_2O): 7% (K = 5.8)

a Which three elements needed by the growing grass did the fertiliser contain?

b Why is it called a 'compound fertiliser 7–7–7'?

c Why do you think it is advisable to treat the soil with a general fertiliser before sowing the grass seed?

d Someone may think that by putting five times as much fertiliser on the soil as is recommended they will get an even better lawn. Give two reasons why adding too much fertiliser may, in fact, lead to a very poor lawn developing.

31 The compost heap

A fourth year class investigated the community of animals living in the school compost heap. They collected some of the compost in a plastic bag. Two of the class set up a Tullgren funnel with 50 g of the compost on the mesh in the funnel. The light above the funnel was switched on and the apparatus was left for four days until the next lesson, when the animals which had collected in the meths were removed.

Six other groups each had 500 g of compost at one end of a tray. Each group sorted through the material systematically, removing any animals found and placing them in a large specimen tube half filled with meths. The groups searched for exactly 20 minutes. When they had finished, the compost was returned to the compost heap.

In the following lesson the class used keys to identify the animals they had collected. They counted the numbers of each type. They then pooled their results, which are shown in the table below.

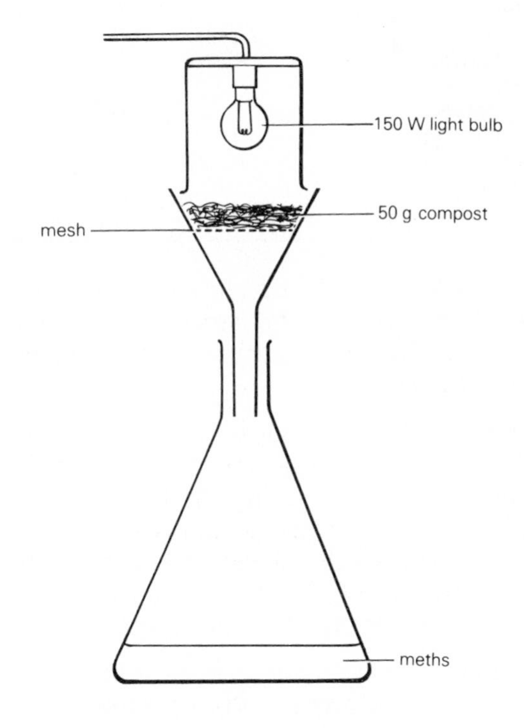

a Explain how the Tullgren funnel works.

b Why is it necessary to collect the animals in meths rather than in water?

c Why was it important that all groups were given the same time for sorting?

d Darren remarked that the Tullgren funnel was much more effective at removing soil organisms, whereas Armajit disagreed.
(i) Give one disadvantage and one advantage of hand-sorting.
(ii) Give one disadvantage and one advantage of using the Tullgren funnel to collect animals from the compost.

Species	**Diet** (H = herbivore; C = carnivore)	Total number of animals collected	
		By hand-sorting	**from Tullgren funnel**
Woodlice	H	15	4
Snails	H	12	–
Centipedes	C	9	3
Spiders	C	23	15
Millipedes	H	18	2
Harvestmen	C	6	–
Slugs	H	8	–
Springtails	H	35	285
Bristletails	H	22	147
Beetle larvae	C	7	–
Nematodes	H	42	2
Leather-jackets	H	3	–
Mites	H	5	62
Maggots	H	11	4
Ground beetles	C	6	–

e Study the table. Draw a new table to show the total numbers of herbivores and the total numbers of carnivores collected by each of the methods.

f Assume that the groups using hand-sorting collected all the animals in their samples, except the springtails, bristletails and mites, and assume that the Tullgren funnel collected all of these three species in the 50 g sample. Calculate the total number of each species per kilogram of compost.

g Draw to scale a diagram to show the numbers of herbivores and carnivores in 1 kg of school compost.

32 New House Farm – a managed ecosystem

New House Farm is a medium-sized farm in West Yorkshire. The farmer and his team manage their ecosystem to produce crops and animals for sale. In 1985, they had to spend a considerable sum of money. They rented 120 acres of land at £35 per acre. They had to buy seeds of barley, wheat and oilseed rape for planting. These seeds cost £6000. The fertiliser bill was £9000 and the chemical sprays cost £5000. The beef calves were bought for £3000, but the biggest item was the wage bill which for the farmer and workers totalled £20 000. Nearly as big was the machinery bill for £15 000 which included both buying new machines and repairs. Diesel to run the tractors cost £5000. The animals cost nothing extra to feed, as they used only food grown on the farm.

In 1985 they harvested 200 tons of barley and sold it at £120 per ton; 110 tons of wheat were sold at £120 per ton and 60 tons of oilseed rape at £300 per ton. They also sold 100 tons of potatoes at £55 per ton. Twenty beef cattle were sold at market for £550 each. They also sold 20 pigs at £40 each and 5 lambs at £20 each. Six turkeys were reared and sold at £6 each, and they sold 200 dozen eggs at 60p per dozen.

Obviously, the farmer's aim is to sell the crops and animals so that the money spent is balanced by the money earned. He hopes to make a profit. The question is, *did* the farm make a profit?

a Draw a table similar to the one below and include *all* the information that is given in the account of New House Farm.

The cost of running the farm		Money from the sale of produce	
ITEM	TOTAL COST	ITEM	TOTAL COST

b From your completed table work out
(i) the total running cost for New House Farm in 1985;
(ii) the total amount of money from sales in 1985.

c How much profit did New House Farm make in 1985?

d Overall, £14 000 was spent on fertilisers and chemical sprays. Explain clearly why this outlay was necessary. What do you think would have happened if the farmer had saved this expense and bought a large car instead?

e They were able to sell 100 tons of potatoes, yet potatoes do not appear as an item of cost for running the farm. Explain this apparent omission in the accounts.

f The best price is obtained for oilseed rape, at £300 per ton. Explain why the farmer is wise not to turn all his fields over to growing this crop instead of barley, wheat and potatoes.

Man and the environment

As the human population grows, man continues to exploit the environment. Some of the effects of this are considered in this group of questions.

33 Fertiliser pollution

When fertilisers such as nitrate and phosphate are used in farming many of them are washed into rivers by rainfall. The rivers carry them into the sea. The Baltic Sea is almost completely surrounded by farming land, and there is very little mixing of its waters with its neighbour, the North Sea.

The graph shows how the concentration of mineral salts dissolved in the Baltic Sea has increased since 1900 and also how the concentration of oxygen dissolved in it has decreased.

a What was the mineral salt concentration in 1930?

b What was the oxygen concentration in 1950?

c Suggest what effect the increased concentration of dissolved mineral salts would have on the population of algae in the sea.

d Suggest how the change in the population of algae has caused the oxygen concentration in the sea to drop so much. (Consider why many algae would die and what would happen to them once they were dead.)

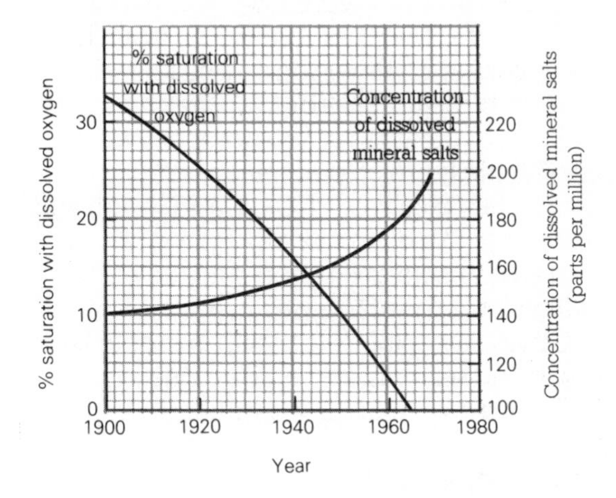

34 Pollution indicators

Many organisms can be used as pollution indicators, that is they give us information about pollution levels without the need for expensive testing equipment. Lichens can be used as indicators of air pollution. They grow mainly on the trunks of trees and on roofs.

The effect of air pollution on lichens was studied in the area around Newcastle-upon-Tyne. The graph shows the number of lichen species found at various distances from the city centre and the amount of sulphur dioxide in the atmosphere at the same points.

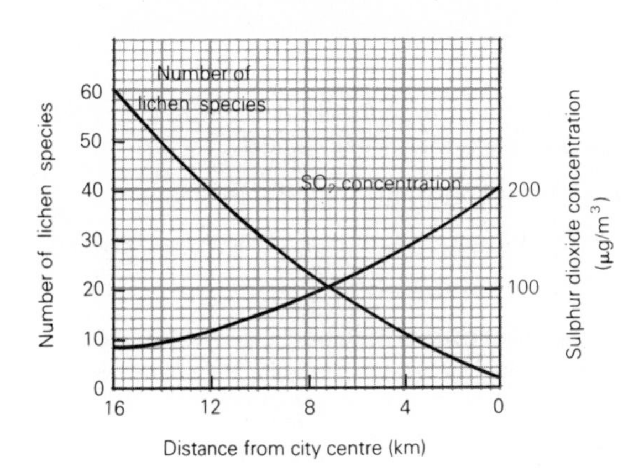

a Describe the relationship between the number of lichen species and the distance from the city centre.

b Give one possible reason for this relationship.

c Design an investigation which would provide evidence for the level of sulphur dioxide in the atmosphere around your school.

35 Whale hunting

Whales are hunted because their meat and the oil extracted from their blubber are very valuable. The table opposite shows the estimated population of blue whales in Antarctica over the last hundred years.

Year	Estimated number of blue whales
1900	50 000
1930	30 000
1950	10 000
1962	1000
1975	5000
1981	16 000

a Draw a graph to show these changes in the blue whale population.

b Describe what has happened to the population of these whales over the last 80 years.

c Suggest reasons for any rises or falls you have described.

36 Culling

In the last century many parts of the highlands and islands of Scotland were cleared of their native animals to make way for sheep grazing. Recently, one of the islands was cleared of sheep, and deer were re-introduced. The deer multiplied rapidly, and soon it became necessary to kill some of them so that there was enough food for the rest. (This process is known as culling.) The meat from the animals which were culled was sold as venison. It was found that the island produced more meat from the culling of the deer than it did when sheep were farmed.

In some parts of Africa it is found that similar results are obtained if the eland (a large antelope) is re-introduced instead of using the land for cattle-ranching.

Explain why culling a natural population is often more successful than the farming of introduced animals such as cattle.

37 Air pollution

The table below shows estimates of the total amounts of various pollutants which were emitted into the atmosphere in the USA in 1970.

Pollutant	Mass (million tonnes)
Carbon monoxide	147
Sulphur oxides	34
Nitrogen oxides	23
Hydrocarbons	35
Particles, e.g. carbon	26
TOTAL	265

a Calculate the percentage of the pollution caused by each of the gases.

b Draw a pie-chart to show the data.

c What would be the major source of each of these pollutants?

d Suggest one damaging effect on human beings or the environment of each of the pollutants.

38 Lead poisoning

The data below show some information on the estimated intake of lead by people in the USA.

a According to this data, how much lead would be absorbed per day into the blood of an average American living in a city and smoking 30 cigarettes per day?

b How much lead would be absorbed into the blood of a non-smoking American who lives in a rural area?

c What proportion of the lead taken into the body is actually absorbed into the blood from
(i) food;
(ii) water;
(iii) air;
(iv) tobacco smoke?

d Suggest possible reasons for the differences in the amounts absorbed.

e What is the major source of the lead pollution in urban air?

Average daily intake of lead

Substances	Average daily intake	Lead concentration	Total lead taken into body (e.g. into gut or lungs) (mg)	Total lead absorbed into blood (mg)
Food	2 kg	165 mg/kg	330	17
Water	1 kg	10 mg/kg	10	1
Urban air	20 m^3	1.3 mg/m^3	26	10.4
Rural air	20 m^3	0.05 mg/m^3	1	0.4
Tobacco smoke	30 cigarettes	0.8 mg/cigarette	24	9.6

39 DDT

Chemicals which kill insects (insecticides) have been used in large quantities all over the world since the 1940s. One of the most widely used has been DDT. In anti-malarial campaigns its use has saved tens of millions of human lives. Its use in killing pests of food crops has saved thousands of millions of pounds.

However, it does have some side effects which have caused its use to be banned in many countries. Many useful insects are killed when it is sprayed. Every animal in the world now has some DDT in its body; DDT accumulates at each stage in a food chain. Many carnivores, such as birds of prey, have almost been wiped out by this accumulation of DDT. On the other hand, many species of insect are now immune to DDT.

The pyramid of numbers below shows the amounts of DDT found in different trophic levels in the organisms of an American estuary in the 1960s.

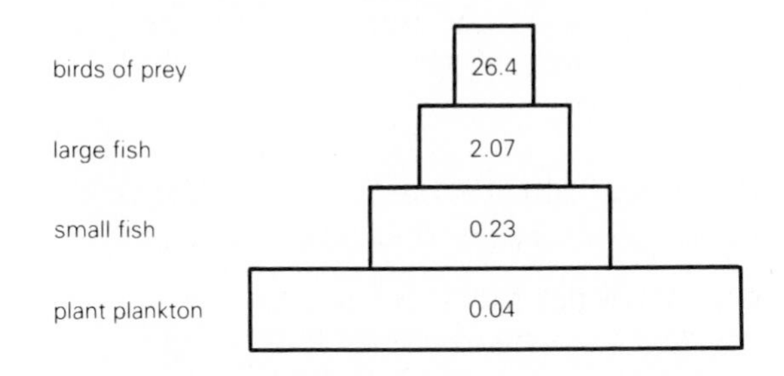

a Explain why the concentration of DDT in an organism is higher than the concentration in its food.

b Calculate by how much the DDT was concentrated at each stage in the food chain.

c The concentration of DDT in the water was only 0.000 05 parts per million. By how much did the plant plankton concentrate the DDT?

The graph below shows the changes in thickness of the eggshells of two birds of prey in Britain between 1920 and 1970. Eggs with thin shells are often broken before they hatch.

DDT was widely used in Britain from the mid-1940s to the early 1960s. Its use was then restricted, and other insecticides which do not persist in the environment for so long were brought into use.

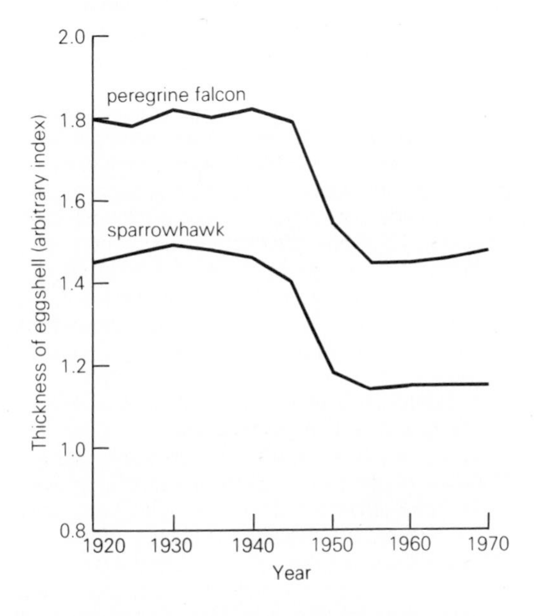

d What evidence do the graphs provide that DDT affects the thickness of the eggshells?

e Suggest why it was the numbers of birds of prey rather than the numbers of smaller birds that were affected most by the introduction of DDT.

f Suggest how you could prove that DDT does affect the thickness of the shells.

g If you were a scientist working for a chemical company which is developing a new insecticide, what factors would you take into account before recommending the company to market it?

40 Biological control

Biological control is the use of a pest's natural enemies to destroy or keep down the numbers of that pest. Greenhouse owners can, for example, use a natural enemy of the whitefly to control this common pest.

Read the following account which is adapted from an article in a gardening magazine.

> The whitefly is an important pest of plants like tomatoes and cucumbers. The adults, which occur in large numbers on the undersides of upper leaves, are small moth-like insects about 1 mm long. They are covered in wax, which gives them a pure white appearance. The eggs, laid mostly on the undersides of the leaves, hatch into minute larvae which disperse before feeding. These turn into flat, semi-transparent immobile 'scales' which feed on plant sap. Each 'scale' changes into a pupa, a non-feeding stage from which the new adult whitefly emerges. The whole life cycle takes about three weeks at 25 °C, longer at lower temperatures. Whitefly can survive freezing conditions for up to a month, so they may survive outside the greenhouse during the winter and infect young plants in early spring.
>
> If whiteflies are allowed to build up to large numbers they will kill plants. Long before that occurs, they excrete a sweet substance known as 'honeydew', upon which sooty mould soon grows. These dark growths make the plant look dirty and cut down the light reaching the leaves. Whiteflies also suck the sap, further weakening the plant.
>
> The whitefly's natural enemy is a parasite called *Encarsia formosa*. The wasp-like adult is about half the size of a whitefly. A female (males are rare) lays about 50 eggs. Each is inserted into a different whitefly scale, and the young parasite feeds on the larva within the scale. After about 9 days at 25 °C the scale turns black, and after a further 11 days the adult parasite, having eaten its host, emerges through a hole in the upper surface. It then flies towards whitefly, attracted by their smell. It can find and parasitise single scales several metres away.

a Using the account above, draw diagrams showing the life cycles of
(i) the whitefly;
(ii) the parasite, *Encarsia*.

b Explain carefully how the plants in a greenhouse would be affected by

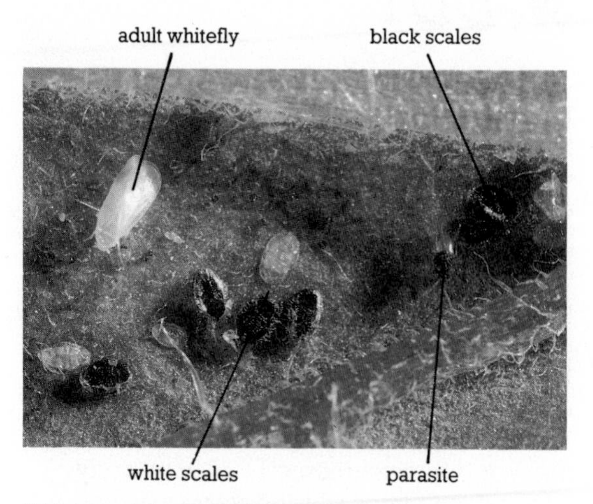

Whitefly and the parasite *Encarsia* on leaf surface

(i) the feeding method of the whiteflies;
(ii) the growth of sooty moulds.

Gardeners can buy *Encarsia* to put in the greenhouse. They buy strips of leaf on which are the black scales of the whitefly which show that the parasite is growing inside the scales. Gardeners are recommended to buy a thousand scales for a 40 m^2 greenhouse, as long as they are put in the greenhouse as soon as the whitefly first appear.

c Suggest why it is best to sell the *Encarsia* in the scales, rather than as eggs or adults.

d A gardener buys 1000 scales. If 80 percent emerge as adults, how many whitefly larvae could be killed?

Whitefly can also be killed with insecticides. Opposite is a table showing the approximate cost of treating a greenhouse with various insecticides.

e Which insecticide would you recommend? Explain your answer.

f The table opposite gives only the approximate cost of treating a greenhouse with insecticides. Suggest three factors which would affect the precise cost of

Insecticide	Cost	% of whitefly killed	
		Adults	Pupae
Dimethoate spray	18p	50–80	10–50
Malathion spray	12p	80–95	less than 10
Permethrin spray	12p	over 95	80–95
Permethrin Smoke Cones	70p	10–50	less than 10
Pyrethrum spray	210p	over 95	50–80

treating a greenhouse with a particular insecticide.

g Suggest why a greenhouse may have to be sprayed several times to get rid of a whitefly infestation.

h Suggest why the insecticides are less effective against the pupae.

i 1000 *Encarsia* cost about £6.00. Give at least two reasons why a greenhouse owner might prefer this method of control.

j Whitefly also affect cabbage and other crops outside the greenhouse. Which method of control would you use to control whitefly on these crops? Explain your answer.

UNIT 6

Structure and chemistry

Most living organisms are made up of cells and contain similar chemical substances, such as proteins, carbohydrates and fats. Many of the chemical processes which go on inside cells are also very similar in all living organisms, and enzymes are always used to control these processes. The questions in this unit are about cells and the chemical processes which go on inside them.

41 Cells

The photographs below show some cells from the leaf of a flowering plant and from the body of a mammal.

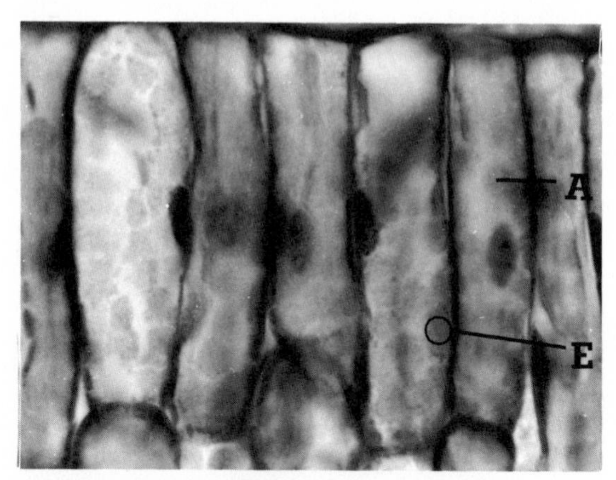

4.7 μm
palisade cells cut across

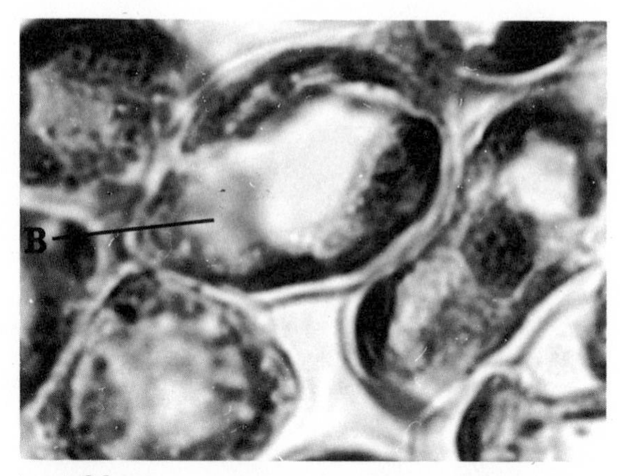

2.1 μm
palisade cells from the leaf of a flowering plant

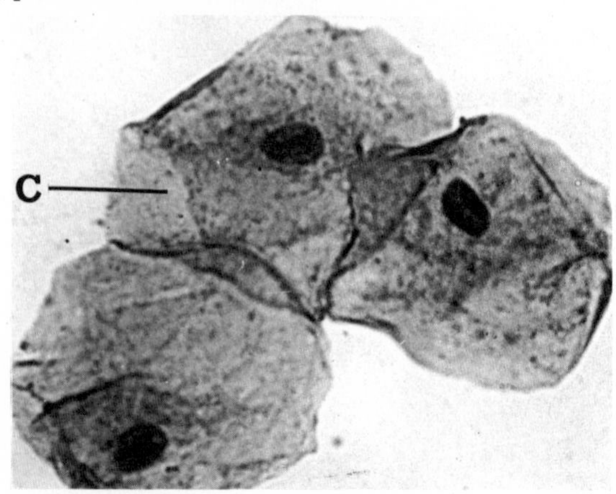

8.9 μm
cells lining the inside of a human cheek

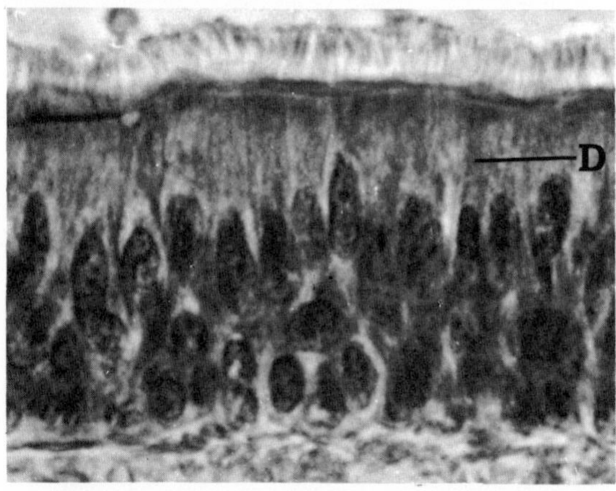

4.3 μm
cells lining the windpipe of a mammal

a Examine the photographs carefully. Give at least four differences between the palisade cells and the cells from the lining of the windpipe.

b Give at least two differences between the cheek and palisade cells.

c Use the scales to calculate the sizes of the following in micrometres:
(i) the length of the palisade cell, A;
(ii) the diameter of the palisade cell, B;
(iii) the width of the cheek cell, C, at its widest point;
(iv) the length of the cell, D, from the lining of the windpipe;
(v) the length of a chloroplast, E.

d Look carefully at photographs A and B and then describe the shape of a palisade cell.

e Suggest one advantage of the spaces between the palisade cells.

f Suggest why there are no spaces between the cells lining the windpipe.

g Draw an accurate diagram of one of the palisade cells in photograph A, and label the parts that can be seen.

h Make a similar labelled diagram of a cell from the lining of a windpipe.

42 Water

The table below shows the approximate percentage of water in different parts of a mammal's body.

Part of the body	Percentage of water
Blood	80
Bone	25
Brain	85
Fat	20
Kidney	80
Liver	70
Muscle	75
Average for whole body	70

a (i) Which part of the body contains most water per 100 g?
(ii) Which part of the body contains least water per 100 g?

b An adult man has a mass of 70 kg. He has 30 kg of muscle, 5 kg of blood, and his brain weighs 1.4 kg. How much water does
(i) his body contain altogether?
(ii) his muscle contain?
(iii) his blood contain?
(iv) his brain contain?

c A 90 kg man is 20 kg overweight due to the storage of extra fat. Would the proportion of water in his body be more or less than that of the 70 kg man? Explain your answer.

d On average a man loses about 2.5 kg of water per day.
(i) Give four ways in which water may be lost from the body.
(ii) How is this lost water replaced each day?
(iii) Suggest three conditions which would cause a man to lose more water per day than average.

e A 2 kg chicken will weigh much less when it is taken out of the oven after roasting. Explain why.

f (i) 200 g of meat, which is mainly muscle, is completely dehydrated. How much would the dehydrated meat weigh?
(ii) Explain why the dehydrated meat would keep much longer than the fresh meat without going bad.

g Below is a list of some of the properties of water.
(i) good solvent
(ii) transparent
(iii) high heat capacity (i.e. a lot of heat

energy is needed to raise its temperature)
(iv) high latent heat of evaporation
(v) solid ice is less dense than liquid water.
Describe one way in which each of these properties is important to living things.

h An apple grower wants to keep her apples as long as possible before selling them. To keep them looking fresh she must prevent them losing water. It is suggested that they can be sprayed with a substance which is impermeable to water.
(i) Describe a test she could do to find out whether the spray really works.
(ii) What information would she need before she could decide whether or not it would be safe and economical to use the spray?

43 What are we made of?

The table below shows the percentage of each of the main elements in the human body.

Element	Percentage (by mass)
Oxygen	65
Carbon	18
Hydrogen	10
Nitrogen	3.5
Calcium	1.5
Phosphorus	1.0
Iron and other mineral elements	1.0

a Using these figures draw a pie chart to show the proportions of different elements in the human body.

The elements are, of course, present in the body in a very large number of compounds. The proportions of the main compounds in the body are approximately:

water	70%
fats	15%
proteins	12%
carbohydrates	0.5%
minerals	2.5%

b Which compounds will contain most of the carbon?

c Which compounds will contain most of the nitrogen?

d Which compound will contain most of the oxygen?

e Explain why the percentage by mass of oxygen is so much higher than the percentage of hydrogen, even though, for example, the chemical formula for water is H_2O.

f Which part of the body will contain most calcium?

g Which part of the body will contain most iron?

44 Food tests

A pupil carried out some tests on five different foods. The tests she used were the iodine test for starch, Benedict's test for sugar, and the biuret test for protein. Her results are shown opposite.

a Prepare a table to show the food substances contained in each of the five foods.

b Which foods contained starch?

Food	Colour after testing with food test reagent		
	Iodine test	Benedict's test	Biuret test
A	black	orange	pale blue
B	yellow	orange	purple
C	yellow	orange	pale blue
D	black	blue	purple
E	black	orange	purple

c Which food contained sugar only?

d Which foods contained both starch and sugar?

e Which food contained starch, sugar and protein?

f Potato contains starch and protein but not sugar. Which food might have been potato?

45 Effects of temperature on an enzyme

A class of pupils measured the time taken for an enzyme to digest starch at different temperatures. Each group in the class kept a mixture of starch and the enzyme at a particular temperature. The mixture was tested for starch every half minute. The time taken for the starch to disappear completely, i.e. to be digested by the enzyme, was noted by each group. The class results are shown in the table below.

a Plot a graph of these results. Use the horizontal axis for temperature and the vertical axis for the time to digest starch.

b From your graph, what is the optimum temperature for the action of this enzyme?

c Explain why the starch takes so long to be digested at 10 °C.

d Explain why starch was not digested at 70 °C.

e If this enzyme were present in the human body, what would you expect to happen to the rate of digestion if the body temperature increased by 3 °C?

f (i) Which group's result was anomalous (that is, did not fit with the pattern obtained by the other groups)?

Group	Temperature (°C)	Time to digest starch (minutes (to nearest half minute))
A	10	25
B	20	9
C	25	7½
D	30	3
E	35	2
F	40	1½
G	45	1½
H	50	2½
I	55	starch not digested in 30 min
J	60	starch not digested in 30 min
K	65	2½
L	70	starch not digested in 30 min

(ii) Suggest a possible explanation for this group's anomalous result.

g (i) Describe how you would carry out the investigation to find the rate of digestion of starch by this enzyme at 50 °C.

(ii) List the apparatus you would need for this investigation.

h How could group A keep the temperature constant at 10 °C?

46 New and old spuds

Cilla and Sally were trying to find out whether old potatoes contained more starch-digesting enzyme than new potatoes. To do this they used dishes containing starch in an agar jelly. They cut out a piece of old potato and a piece of new potato, and put each onto a starch/agar dish, as shown below.

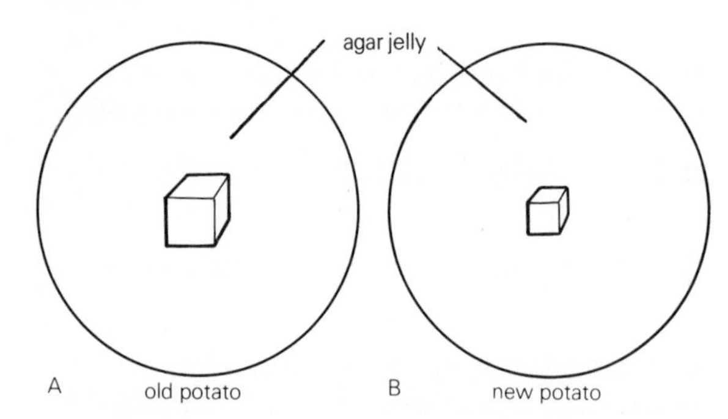

After 24 hours they removed the pieces of potato and poured iodine solution on the starch/agar. The results are shown below.

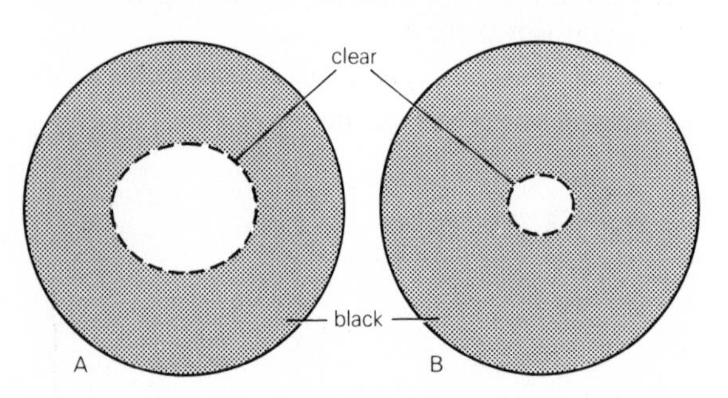

dishes after flooding with iodine solution

They concluded that old potatoes contain more starch-digesting enzymes than new potatoes.

a List the precautions Cilla and Sally should have taken to make sure the test was absolutely fair.

b Explain why there were clear areas in the starch/agar after iodine solution was poured onto it.

c (i) What substance, other than agar, would you expect to find in the clear area on plate A?
(ii) How could you test the agar for this substance?

d Was Cilla and Sally's conclusion justified? Explain your answer.

e Suggest why old potatoes might be expected to contain starch-digesting enzyme, whereas new potatoes would not.

47 Cheese-making

Rennin is an enzyme which causes milk to clot. It is used in cheese-making to start making the curd.

Below are the results of an experiment in which the time taken for milk to clot after adding rennin was measured at different temperatures.

Temperature (°C)	Time to clot (min)
15	no clot
20	35
25	18
30	8
35	5
40	3
45	2
50	5
55	7
60	no clot

a Plot a graph of these results.

b At what temperature was clotting fastest?

c From these results, what would be the best temperature for making curd?

d What other factor might a cheese-maker take into account when deciding the best temperature to use for making curd?

48 Brown potatoes

Below are the results of an investigation with a colourless substance called catechol and juice extracted from potato. The colourless catechol can be changed into a brown substance.

a What conclusion about potato juice can you draw from the results in tubes A and B?

b What evidence is there that an enzyme is involved in the browning process?

c What result would you expect if the temperature of the tubes were raised from room temperature to 40 °C?

d Which tubes could be described as controls? Explain your answer.

e Suggest why the potato juice in tube D slowly turned brown.

f (i) Pieces of cut potato go brown quite quickly. If the cut surfaces are covered with cling-film the browning is much slower. What does this suggest about the browning process?
(ii) Describe an experiment to test your suggestion.

g Describe an experiment to find out whether bananas also contain a substance which makes catechol go brown.

Tube	Contents of tube	Result
A	Catechol and potato juice	rapid browning of catechol
B	Catechol only	no browning
C	Catechol and boiled potato juice	no browning
D	Potato juice only	slow browning

A cookery book says that before freezing new potatoes, they should be 'blanched'. The instructions for blanching the potatoes are:

1. Bring a large pan of water to boil.
2. Plunge in the basket of potatoes, cover and bring quickly to boil again.
3. Leave the potatoes in the boiling water for four minutes.
4. Remove the basket of potatoes and quickly plunge it into ice-cold water, until the potatoes are completely cold.
5. Drain thoroughly before freezing.

h Four minutes is not long enough to cook potatoes. Suggest why the potatoes should be boiled for four minutes before freezing.

i Suggest why the potates are cooled quickly after boiling.

j The same book suggests that peas should also be blanched in the same way before freezing. However the time given for boiling is only 30 seconds. Explain why.

UNIT 7 Plant nutrition

Green plants have the unique ability to make their own food from simple inorganic substances, using the energy from sunlight. This process is photosynthesis, and all other living organisms depend for their food on the substances that green plants make in photosynthesis. The questions in this unit are about photosynthesis, and about how we can make use of our knowledge of photosynthesis to increase food production.

49 Leaves

A plant with green and white leaves was destarched. Part of one leaf was then covered on both sides with black paper as shown. The plant was left in bright light for six hours. The leaf was then removed from the plant and tested for starch.

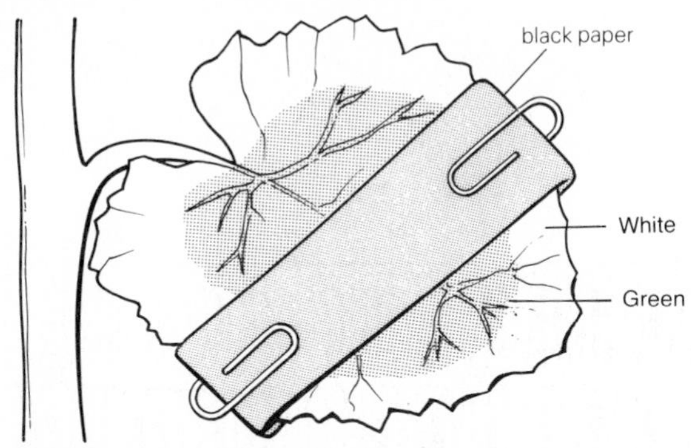

a Explain
(i) how you would destarch the plant;
(ii) why this is necessary.

b How would you dissolve out the green chlorophyll before testing the leaf for starch?

c Draw a diagram of the leaf as it would appear after testing for starch with iodine solution. Label the colours you would see in different parts of the leaf.

Below is a simplified diagram of a small section through the green part of a leaf.

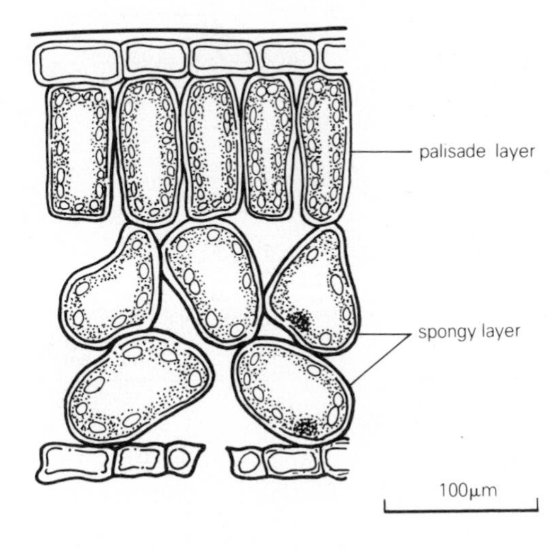

d Count the number of chloroplasts that you can see in *each* of the cells of the palisade and spongy layers. Record your results in a table which shows
(i) the number of chloroplasts in each cell in each layer;
(ii) the total number of chloroplasts in each layer;
(iii) the average number of chloroplasts per cell in each layer.

e Suggest a reason for the difference between the numbers of chloroplasts in the two layers.

f The scale on the diagram shows a length of 100 micrometres (one micrometre is one thousandth of a millimetre). Use the scale to find:
(i) the thickness of the leaf;
(ii) the thickness of each of the four cell layers.

g Explain the advantages for photosynthesis of leaves being so thin.

50 Temperature and photosynthesis

The graph below shows the results of an investigation into the effect of temperature on the rate of photosynthesis.

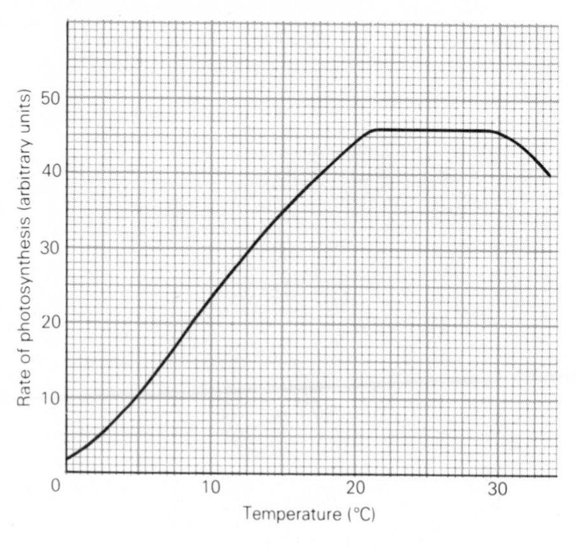

a What was the rate of photosynthesis at 10 °C?

b Over what range of temperatures was the rate of photosynthesis at its fastest?

c What conclusions can you draw about the effect of temperature on photosynthesis? Consider the full range of temperatures shown on the graph.

d A greenhouse owner wishes to grow lettuces as rapidly and economically as possible in winter. What advice would you give about the best temperature at which to keep the greenhouse? Explain your answer.

e What else might the greenhouse owner need to provide for rapid growth of the lettuces?

51 Stomata

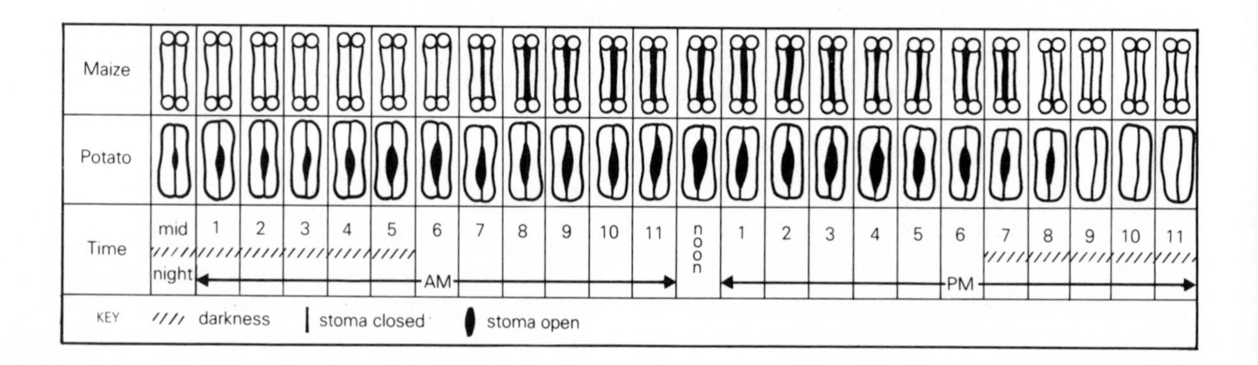

The diagram shows the daily opening and closing of the stomata in two plants, the potato and maize.

a At what time did
(i) the maize stomata open?
(ii) the maize stomata close?

(iii) the potato stomata open?
(iv) the potato stomata close?

b (i) In which plant were the stomata open for the longer time?
(ii) How many hours longer were the stomata open in this plant?

c How many hours of darkness were there?

d Textbooks sometimes say that stomata are open during the day and are closed at night. Is this hypothesis supported by the information on potato and maize? Explain your answer.

52 How do stomata open and close?

The diagrams below show sectional views through a stoma and the surrounding guard cells.

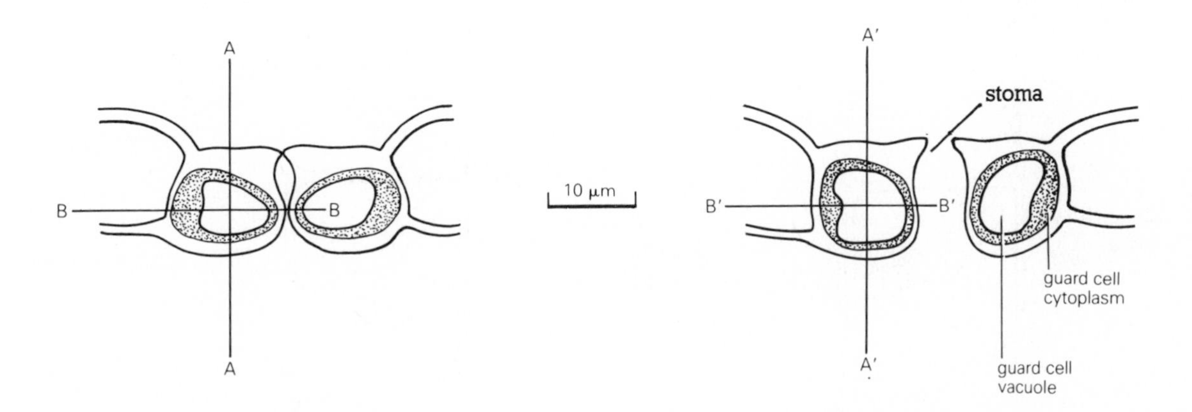

a Calculate the width of the open stomatal pore.

b Draw a table like the one shown below. Use the scale on the diagram to work out the measurements and fill in the table.

	Stoma closed (μm)	Stoma open (μm)
Depth of guard cell along AA or A′A′		
Width of guard cell along BB or B′B′		
Depth of vacuole along AA or A′A′		
Width of vacuole along BB or B′B′		

c The opening and closing of the stoma is thought to be caused by changes in the pressure inside the guard cells due to osmosis. Using only the evidence in the diagrams and the information in your table, describe whether the guard cells are turgid, that is filled with water as a result of osmosis,
(i) when the stoma is closed;
(ii) when it is open.

d Suggest how osmosis may cause the stomata to open.

53 Oxygen production in photosynthesis

In the 1880s a German biologist called Englemann carried out a number of ingenious experiments on photosynthesis. He used the freshwater alga called *Spirogyra* and a freshwater bacterium which can swim and which moves into regions of high oxygen concentration. The diagrams below show what he saw when he (A) kept the *Spirogyra* and bacteria in bright light, and (B) shone two small spots of light on the *Spirogyra*.

a What do you notice about the distribution of the bacteria
(i) in the light?
(ii) in the dark with two spots of light?

b Which part of the *Spirogyra* cell is responsible for producing oxygen? Explain how you worked this out.

c 'Oxygen is only produced in the light when a plant is photosynthesising.' Explain how Englemann's experiment shows this.

d (i) Design an experiment using Englemann's basic idea to find out whether red, yellow, green or blue light is the best for photosynthesis.
(ii) What measurements would you make?
(iii) How could you use your results to help someone improve their production of tomatoes in a greenhouse?

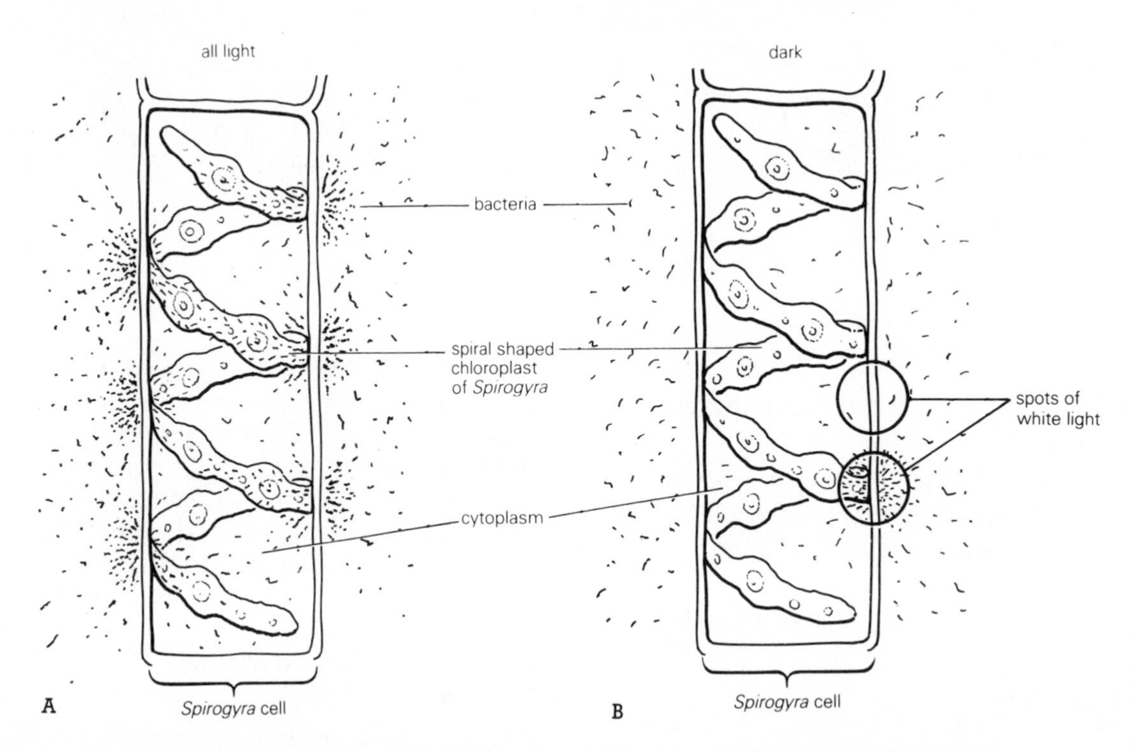

54 Light intensity and photosynthesis

A group of students wanted to compare how plants that live in shade and plants that live in an open meadow respond to changes in light intensity. They made use of the fact that when a leaf of a land plant is placed in water it will draw in water through its stomata if it is respiring only; if it is photosynthesising it will force water out or expand slightly. They used the apparatus shown opposite.

The apparatus was assembled

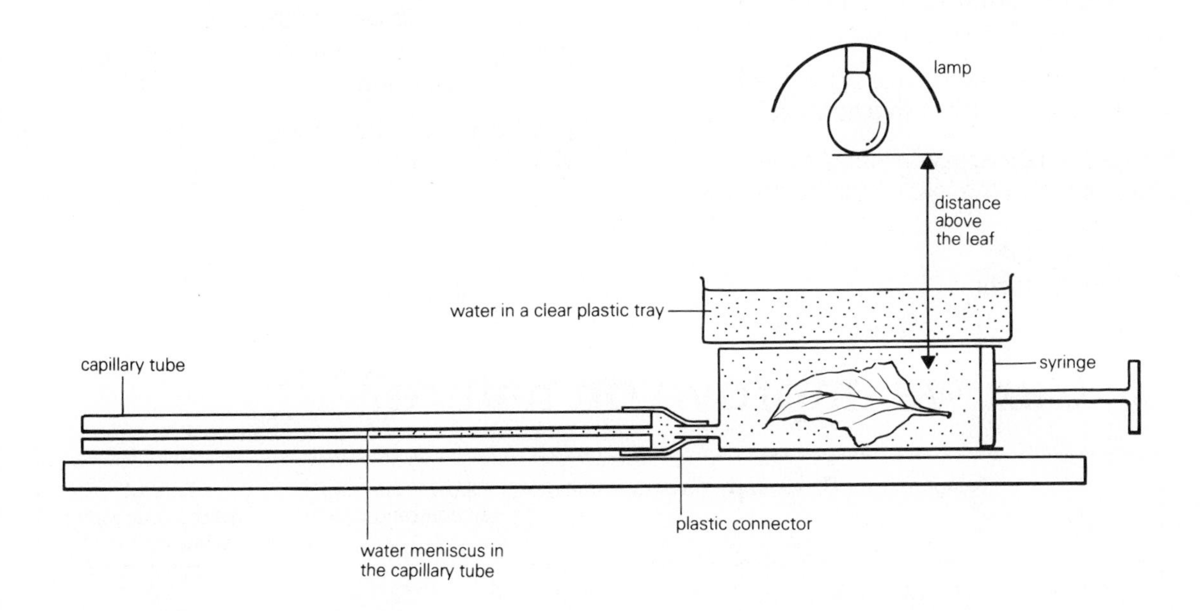

underwater and then taken out. When it was on the laboratory bench a gentle pull on the syringe plunger drew air half way along the capillary tube. During the experiment the students measured how far the water meniscus moved in five minutes. They started with the syringe covered in black polythene and made a measurement of the meniscus movement at five minutes. Next the lamp was placed at the maximum distance from the syringe and the leaf was illuminated. The meniscus movement was measured for a five minute period. The lamp was moved closer and the meniscus movement over five minutes was measured each time the lamp was moved closer. At the end of the experiment the leaf was removed, carefully dried with soft tissue paper and then weighed. The students then worked out how many millimetres the meniscus moved in five minutes for each gram of leaf.

Below is a set of results for two plants, the stinging nettle and the bramble.

Light intensity (arbitrary units)	Meniscus movement (mm per gram of leaf per 5 min)	
	Bramble	Stinging nettle
0	−19	−46
2	− 9	−33
4	− 5	−27
8	+ 1	−16
12	+ 3	− 6
30	+11	+13
42	+15	+40
55	+14	+50

(− means the meniscus moved right; + means the meniscus moved left)

a Why was the apparatus assembled under water?

b Why was capillary tubing used to measure the movement of the water?

c Why did the experiment start with the meniscus in the middle of the capillary tube?

d What would be a suitable control for this experiment?

e What was the water in the clear plastic tray above the syringe for?

f Why were the results expressed in mm *per gram of leaf* per five minutes?

g Explain clearly what must have happened in the leaf when the meniscus moved
(i) to the left;
(ii) to the right.

h Draw a graph of the results with light intensity on the x-axis and meniscus movement on the y-axis.

i The compensation point is the level of light intensity at which the rate of respiration is exactly balanced by the rate of photosynthesis. From your graph, work out the compensation points of the two plants.

j Which plant is better able to live in shade? Explain your answer.

55 Can we improve on nature?

	Temperature (°C)	Light (arbitrary units)	Carbon dioxide concentration (%)	Rate of photosynthesis (arbitrary units)
A	20	3	0.03	75
B	20	6	0.03	75
C	20	3	0.13	150
D	20	6	0.13	195
E	30	3	0.03	75
F	30	6	0.03	75
G	30	3	0.13	180
H	30	6	0.13	270

The table above shows the rate of photosynthesis under eight different sets of conditions, A to H.

a Which set of conditions gives the fastest rate of photosynthesis?

b What is the effect of
(i) increasing the temperature when there is plenty of light and carbon dioxide?
(ii) increasing the light when there is plenty of carbon dioxide and a high temperature?
(iii) increasing the carbon dioxide concentration when there is plenty of light and a high temperature?

c 0.03 percent is the normal amount of carbon dioxide in the air. What do these results tell you about the benefits of adding carbon dioxide to the air?

d Explain why the rate of photosynthesis is the same in A, B, E and F.

The next table shows the effect on crop yields of adding extra carbon dioxide to a greenhouse.

Crop	Yield: Normal air	Yield: Air with extra CO_2
Lettuces	0.9 kg	1.1 kg (mass of 10 lettuces)
Tomatoes	4.4 kg	6.4 kg (mass of each plant)

e What is the effect on the crop yield of increasing the concentration of carbon dioxide in the air?

f The following have all been suggested as possible ways of increasing the carbon dioxide concentration in a greenhouse. Which do you think would be the best way? Explain your reasoning.
(i) Getting men to work in the greenhouse and breathe out the carbon dioxide.
(ii) Keeping rabbits in the greenhouse.
(iii) Adding extra manure to the soil to encourage more worms and micro-organisms.
(iv) Using paraffin heaters in the greenhouse, because when a fuel burns carbon dioxide and water are made.
(v) Dripping acid onto a heap of limestone or chalk, which is calcium carbonate.
(vi) Releasing carbon dioxide from cylinders of the gas.

56 The 'Green Revolution'

The International Rice Research Institute at Los Banos in the Philippines set out to produce a dwarf strain of rice which would increase rice yields. They crossbred around ten thousand strains of rice and eventually produced strain IR-8 which used fertiliser more efficiently. It also ripened in 120 days whereas most varieties take 150 to 180 days to ripen.

Study the graph, which shows the rice yields in three countries between 1950 and 1968.

a What was the yield of rice for each of the countries in 1955?

b From the graph estimate the rice yield in each of the countries in 1962.

c Strain IR-8 was introduced into West Pakistan and Ceylon (now Sri Lanka) at about the same time. From the graph, when would you estimate IR-8 was introduced?

The productivity success of Taiwan was a result of using strain IR-8, multiple cropping (three crops per year) and increased use of fertilisers. IR-8 cannot stand flooding or prolonged submersion. It requires more fertiliser than other varieties; for example, it needed about five times as much as the traditional rice varieties grown in Pakistan.

d What would be the main problems for farmers in Pakistan and Sri Lanka when they introduced the strain, IR-8?

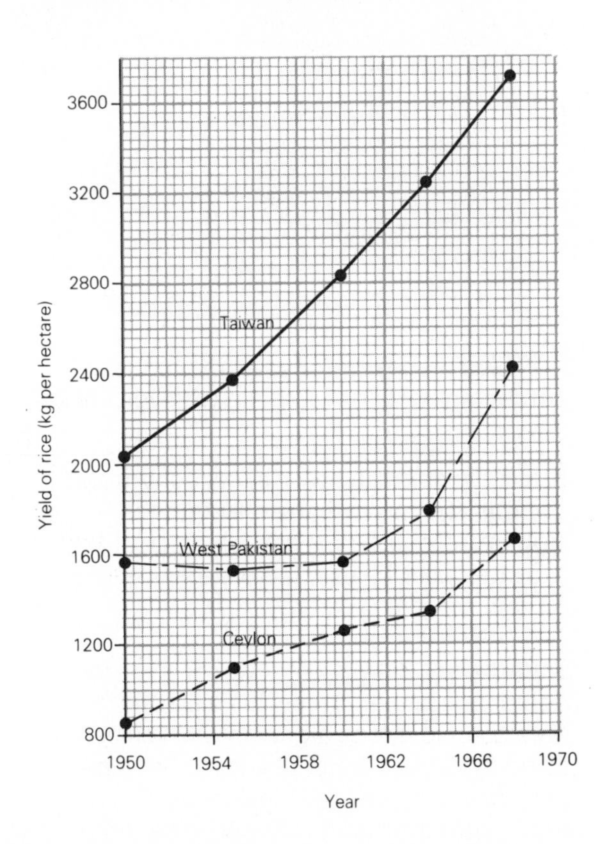

57 Food crops

The following diagram shows the composition of five crop plants in terms of the percentage of carbohydrate, fat and protein.

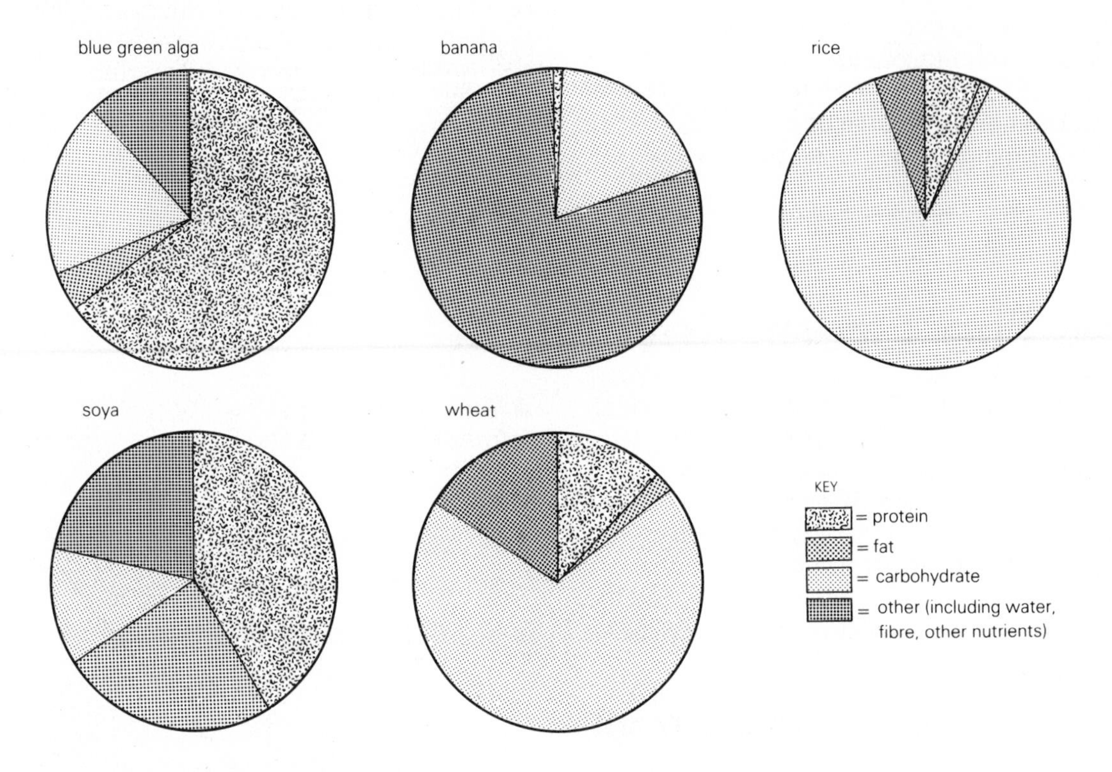

a Which crop does not contain fat?

b Which crop probably has the highest percentage of fibre? Explain why you chose this crop.

c List the crops in order of carbohydrate content, starting with the one with the highest percentage of carbohydrate. Make two other lists to show the orders of fat and protein content.

Blue-green algae grow in warm, mineral-rich ponds in the tropics. In Chad and Mexico blue-green algae are harvested and dried to form a biscuit-like food. Because blue-green algae are single-celled organisms, the food produced is called Single-Cell Protein (SCP).

d What advantages does the blue-green algae crop have over the other four?

e (i) What do you think is the main problem when using SCP as food for human beings?
(ii) How do you think the problem can be overcome?

f SCP is often used to feed animals. Why is it more efficient to feed human beings with it directly?

g Why is it important that we are prepared to research the development of foods for human use like SCP?

UNIT 8

Animal nutrition

Animals get their food from the high energy compounds made by plants, either directly by eating the plants or indirectly by eating other animals. The questions in this unit are about how animals, and in particular human beings, obtain and process their food.

58 Energy from sweets

The table below shows the amount of energy contained in various sweets made by the Mars company. The table is reprinted from a booklet designed for slimmers. Like many slimmers' publications, it shows energy values in calories. 1 calorie is the same as 4.2 kilojoules.

MARS product	Calories (per 25 g unless otherwise stated)
Bounty	117
Galaxy, 70 g	363
Lockets, 43 g	151
Maltesers	120
Marathon	120
Mars, 62.5 g	270
Milky Way	116
Minstrels, 49 g	207
Opals, 51 g	181
Revels	176
Ripple	154
Topic, 53 g	232
Treets, 42 g	248
Tunes, 37 g	136
Twix	120

a Work out the amount of energy in kilojoules for each type of sweet.

b Put the sweets in order of energy value per gram (be careful – some are shown 'per 25 g', others for whole bars).

c (i) Which sweet would be least fattening per gram?
(ii) Which sweet would be most fattening per gram?

d A Bounty bar weighs 60 g. What is its energy value in kilojoules?

e While sitting watching the TV, a boy uses about 6 kJ per minute. How long would it take to use all the energy obtained from the Bounty bar?

f List as many ways as you can think of in which his body will use up energy while he is sitting watching TV.

g While dancing he uses on average 30 kJ per minute. How long would it now take him to use up the energy obtained from a Bounty bar?

h Why does he need so much energy for dancing?

59 Milk

The table below shows the composition of human and cow's milk, per 100 g.

	Mass per 100 g Human milk	Cow's milk
Water	87.1 g	87.6 g
Proteins	1.3 g	3.3 g
Fat	4.1 g	3.8 g
Milk sugar (lactose)	7.2 g	4.7 g
Calcium	34 mg	120 mg
Iron	0.07 mg	0.05 mg

a Which substances does cow's milk contain in a much higher proportion than human milk?

b Which substance does human milk contain in a much higher proportion than cow's milk?

c How much protein is there in a litre (1000 g) of human milk?

d How much protein is there in a litre of cow's milk?

e How much iron is there in half a litre of cow's milk?

f How much cow's milk would you have to drink in order to obtain 1 g of calcium?

A 7 kg infant requires about 14 g of protein per day, and 500 mg of calcium.

g If a mother is able to supply 1200 g of milk per day, does the infant get enough
(i) protein?
(ii) calcium?

h How much human milk would the infant need to drink in order to get enough protein and calcium to satisfy its needs?

i A 7 kg infant also needs 5 mg of iron per day. How much human milk would the baby have to drink in order to get enough iron?

j A mother can only provide 1 to 1.5 kg of milk per day. How can the infant be provided with all the iron it needs?

60 Square meals

On the next page is an analysis of all the food eaten by a 15-year-old girl during the course of a day.

a Use the table of information to find:
(i) which food provided her with the most protein;
(ii) which food provided her with the most fat;
(iii) which food provided her with the most calcium;
(iv) which food provided her with the most vitamin C;
(v) which food contained carbohydrate only;
(vi) how much energy was provided by her breakfast.

b (i) List the items in her diet which were entirely, or almost entirely, obtained from plants.
(ii) Calculate how much protein she obtained from these plant products.

c It is suggested that a balanced diet should contain carbohydrate, fat and protein roughly in the ratio 5:1:1. Is the girl's diet balanced in this way? Explain your answer.

Meal	Food	Mass (g)	Energy content (kJ)	Carbohydrate (g)	Fat (g)	Protein (g)	Calcium (mg)	Iron (mg)	Vitamin C (mg)
Breakfast	cornflakes	25	390	21.3	0.4	2.2	1	1.7	0
	milk	100	270	4.6	3.9	3.2	103	0.1	1.5
	egg	60	370	0	6.5	7.4	31	1.2	0
	bacon	40	560	0	12.6	5.6	3	0.24	0
	bread	30	290	14.6	0.5	2.5	32	0.5	0
	butter	8	240	0	6.6	0	1	0.02	0
Lunch	fish fingers	110	820	17.7	8.3	13.8	47	0.8	0
	ketchup	15	60	3.6	0	0.3	4	0.2	0
	chips	130	1 280	44.2	13.3	4.7	18	1.1	13
	fruit yoghurt	150	570	27.0	1.0	6.1	225	0.15	1.0
Tea	steak and kidney pie	160	1 830	35.5	27.4	14.9	75	2.9	0
	chips	180	1 770	61.2	18.4	6.5	25	1.5	18
	baked beans	150	520	22.7	0.9	7.2	72	2.1	0
	chocolate cake	100	2 100	53.1	30.9	5.8	130	1.6	0
Drinks	instant coffee (5 cups)	20	80	2.2	0	2.9	28	0.9	0
	milk (in drinks)	125	340	5.8	4.9	4.0	129	0.12	1.9
	sugar (in coffee)	25	420	25.0	0	0	0	0	0
	canned drinks	600	1 000	60.0	0	0	24	0.6	0
Other snacks	chocolate	50	1 110	29.7	15.2	4.2	110	0.8	0
	biscuits	60	1 160	44.9	10.0	4.0	72	1.3	0
	banana	130	420	25.0	0	1.4	9	0.5	13
	orange	100	150	8.5	0	0.8	41	0.3	50
TOTALS		2 368	15 750	506.6	160.8	97.5	1 180	18.63	98.4

Official reports suggest the following as the amounts of nutrients required daily by different people.

Person	Total energy (kj)	Protein (g)	Calcium (mg)	Iron (mg)	Vitamin C (mg)
15-year-old boy	12 100	37	700	9–18	30
15-year-old girl	10 400	31	700	12–24	30
Adult man	12 600	37	500	5–9	30
Adult woman	9 200	29	500	14–28	30

d (i) For the 15-year-old girl, how much more energy was in her day's food than she required?
(ii) What effect would this have if she had similar amounts of food every day?
(iii) Which nutrient does she have in approximately the correct amount?
(iv) Suggest how the girl should change her diet to be more suitable for her needs; for example, what foods should she cut out or reduce?

e Suggest three possible reasons why boys and men need more energy per day than girls and women.

f Explain why the 15-year-olds have a higher calcium requirement than the adults.

g Explain why females have a higher iron requirement than males.

61 A hole in the stomach

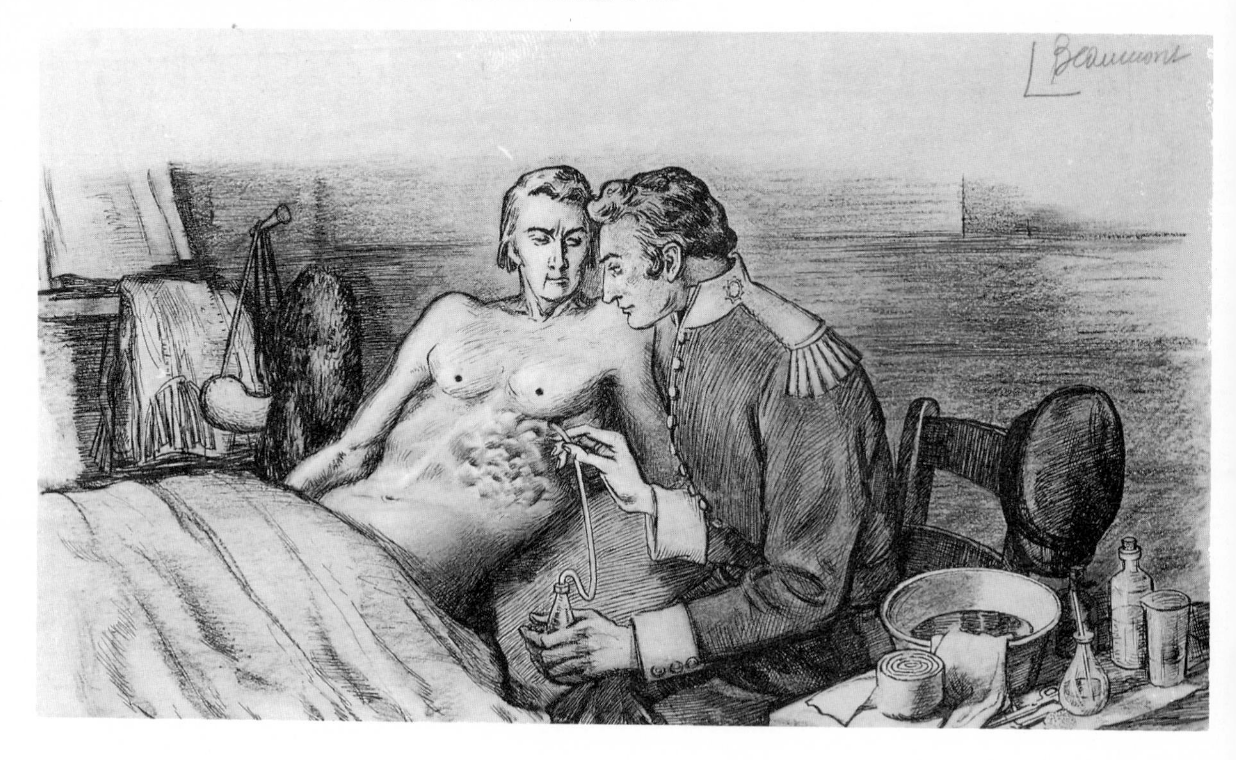

The year was 1822. A party of fur trappers was hunting in a forest in Canada. One of the group tripped and his shotgun accidentally went off. The explosive ripped through the side of the young Alexis St. Martin, tearing a great gash below his ribs and opening his stomach to the outside world. Luckily for Alexis an army fort was nearby and one of the army surgeons was able to treat the young man.

For two years the army doctor looked after his patient. The wound healed slowly, leaving a small hole through his side into his stomach, which the surgeon covered with an artificial flap. This doctor, called William Beaumont, had a scientific turn of mind, and he realised that he had a unique opportunity to study what happened to food inside the human stomach. Over the two years he did over 60 experiments on his patient.

In one experiment he tied lumps of food to a silk thread and then gently pushed them through the hole into Alexis' stomach. Each hour he pulled them out again to see what the stomach juices had done to them. He found that boiled beef was completely digested in less than two hours, but that raw beef and spiced beef were still not digested after 3 hours.

In another experiment he made Alexis starve for 17 hours and then he removed some of the juices through the hole in the stomach. He put the juices in a test-tube, added a lump of boiled beef and put the tube in a saucepan of water which he kept warm at 37 °C. He carefully recorded what happened. Below is a summary of his observations.

Time	Observation
0 minutes	Boiled beef put into tube
40 minutes	Outside of beef digesting. Liquid becoming cloudy.
1 hour	Outside starting to separate.
2 hours	Fibres of meat separated, floating and very soft
4 hours	Only half as many fibres to be seen
6 hours	A few fibres only remaining
8 hours	A few very small fibres floating on surface
10 hours	No sign of fibres. All digestion completed, leaving just a cloudy liquid.

a (i) How long did the boiled beef take to digest completely inside Alexis' stomach?
(ii) How long did the boiled beef take to digest completely in the stomach juices in the test-tube?
(iii) Suggest why there was a difference in times taken.

b Why did Beaumont keep the test-tube in water at 37 °C?

c What control experiment should Beaumont have done?

d What conclusion do you think Beaumont came to about the effects of the stomach juices?

e What other tests could he have done to check these conclusions?

f (i) Can you suggest why the spiced beef might be slower to be digested than the boiled beef?
(ii) How could you test your suggestion?

g Do you think the doctor was right to experiment on his patient in these ways? Explain your views.

62 On the farm

The table below shows how efficiently different farm animals convert their food to meat which can be eaten by humans.

	% total energy in animal's food converted to energy in meat	Protein production (kg per hectare)
Cattle	4	53
Chicken	12	135
Pig	17	105
Rabbit	9	292
Sheep	5	65

a Which type of farm animal is most efficient at converting its food to meat?

b Which type of farm animal is least efficient at converting its food supply to meat?

c How would you expect these differences in efficiency to affect the price of meat?

d Suggest two reasons why one type of animal may be more efficient than another at converting its food supply to meat.

The table also shows how much protein is produced by the animals when they are allowed to graze and feed freely on one hectare of land.

e Which type of animal produces most protein per hectare?

f Suggest why this type of animal is not also the most efficient converter of its food.

g List four other factors which a farmer would take into account when deciding which animals to rear on his land.

63 Digestive enzymes

The diagram below shows the typical pH of samples of the digestive juices taken from the human digestive system at various points.

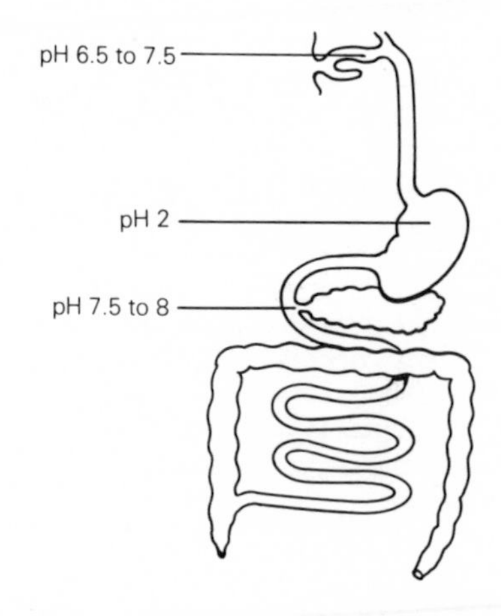

a Draw a table like the one below to show whether each part of the digestive system is acid, alkaline or neutral.

Part	pH	Acid/alkaline/neutral
Mouth		
Stomach		
Small intestine		

b The graph below shows the range of pH over which three enzymes can work. For each enzyme say
(i) at what pH it works best;
(ii) in which part of the digestive system it would work best.

c The three enzymes are:
amylase – which is secreted by the salivary glands,
gastric proteinase – which is secreted by the stomach, and
pancreatic proteinase – which is secreted by the pancreas.
From this information work out which of the enzymes A, B and C, shown on the graph, is which.

d Amylase digests starch into maltose sugar. It is produced by the pancreas as well as by the salivary glands. Suggest why this is necessary.

e Suggest an advantage of the stomach contents being at pH 2.

f Gastric proteinase is made in the cells of the stomach wall in an inactive form. The enzyme only becomes active after it has been secreted into the stomach acid. Suggest an advantage of this system.

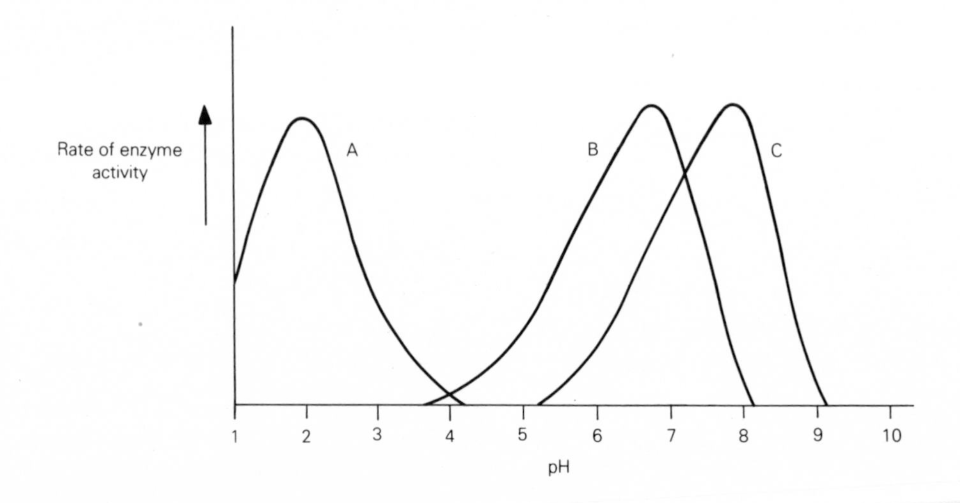

64 The dangers of a sweet tooth

The graph shows the average number of decayed teeth in a group of young children, compared with the amount of sweets the children ate each day.

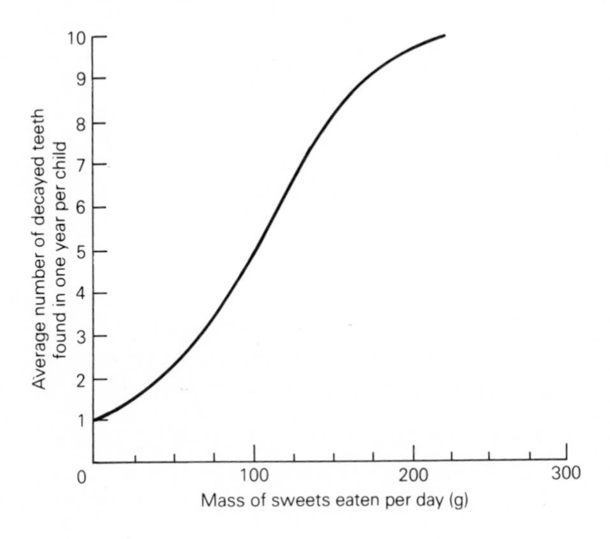

a On average how many decayed teeth were found in children who did not eat sweets?

b On average how many decayed teeth were found in children who ate 200 g of sweets per day?

c What is the relationship between the amount of sweets eaten and the number of decayed teeth found in a child?

d Suggest why decay is more common when sugary foods like sweets are eaten.

e In an investigation into tooth decay, two groups of rats were fed on a sugary diet. One group was normal laboratory rats, but the other group contained rats which had been kept in completely sterile conditions and had no bacteria in their mouths. The normal rats developed tooth decay, but no decay occurred in the bacteria-free rats. Explain why the second group did not develop tooth decay.

65 Two toothpastes

Two types of toothpaste were smeared onto a microscope slide to make a very thin film. At a magnification of ×100 the toothpastes looked like this:

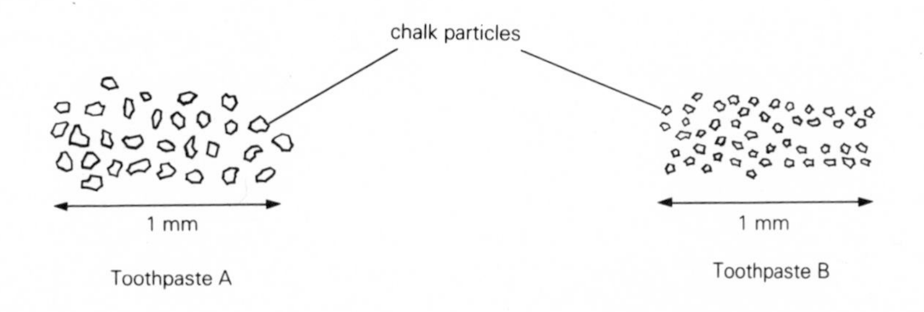

a What differences can you see between the chalk particles in these two toothpastes?

b Estimate the average diameter of the particles in each type of toothpaste. (*Hint*: how many particles would fit across 1 mm?)

c What do you think is the function of the chalk particles in the toothpaste?

d One of these toothpastes is claimed to be especially for 'sensitive teeth'. Which toothpaste do you think this is? What features of the toothpaste might make it better for sensitive teeth?

e What else would you expect to be contained in a toothpaste as well as chalk?

66 The fluoride problem

For many years people have been arguing about whether fluoride should be added to water supplies. Fluoride occurs naturally in the water in some places and is known to be important for making strong bones and teeth. The graph shows the results of measuring the amount of tooth decay in children from places which had different amounts of fluoride in the water. The line has been drawn to show the pattern.

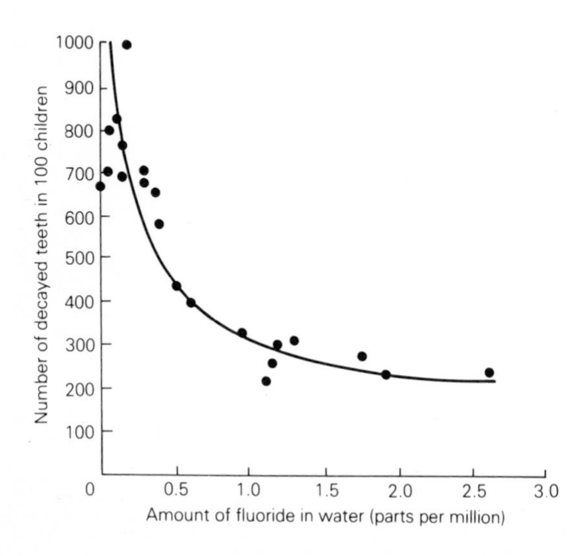

a What happens to the number of decayed teeth per child as the concentration of fluoride is increased?

b About how many decayed teeth does the average child have when there are 2.0 parts per million of fluoride in the water?

c About how many decayed teeth does the average child have when there are less than 0.2 parts per million of fluoride in the water?

d From the evidence in the graph, about how much fluoride would it be most economical to add to the water in places with no fluoride in order to reduce tooth decay in children?

e Many people are opposed to adding fluoride to water supplies. Suggest at least three reasons why it might not be desirable to add fluoride to water supplies.

f What evidence about the safety of fluoride could be obtained by comparing people living in areas with high and low fluoride contents in the water?

g Suggest ways in which children could be supplied with fluoride apart from adding it to water supplies.

UNIT 9 Transport systems

All living cells need certain substances such as oxygen and food from outside the cell. In single-celled organisms these substances can diffuse in and quickly reach all parts of the cell. For larger organisms, however, where the distance from the outside to the cells in the centre is quite large, diffusion is too slow. Larger organisms therefore have transport systems to move substances around their bodies to all cells. The questions in this unit are about the various ways in which substances are moved around organisms.

67 Diffusion

In an investigation to measure the speed of diffusion, coloured gelatin cubes of different sizes were placed in dilute acid solution as shown in the diagram.

As the acid diffused into the cubes, they changed colour. The time taken for each cube to change colour completely is shown in the table.

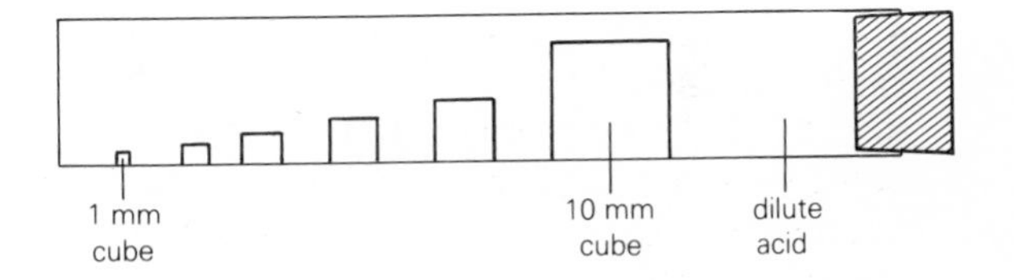

a Copy and complete the table below. The values for the 1 mm and 2 mm cubes have been completed for you.

Length of side of cube (mm)	Surface area of cube (mm^2)	Volume of cube (mm^3)	Surface-area-to-volume ratio
1	6	1	6
2	24	8	3
3			
4			
5			
10			

b Plot a graph to show the relationship between the surface-area-to-volume ratio and the time taken for the cube to change colour completely. Use the vertical axis for time and the horizontal axis for the surface-area-to-volume ratio.

c Describe the relationship between surface-area-to-volume ratio and time taken to change colour.

Length of side of cube (mm)	Time taken to change colour completely (seconds)
1	20
2	42
3	76
4	102
5	180
10	560

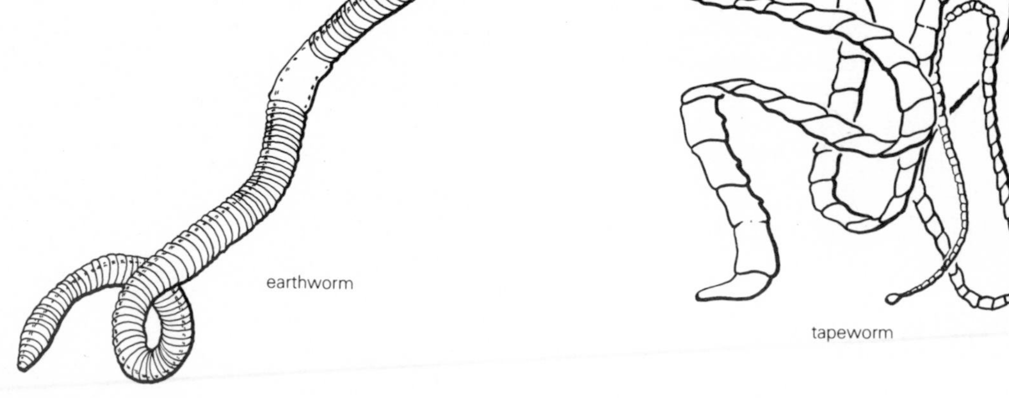

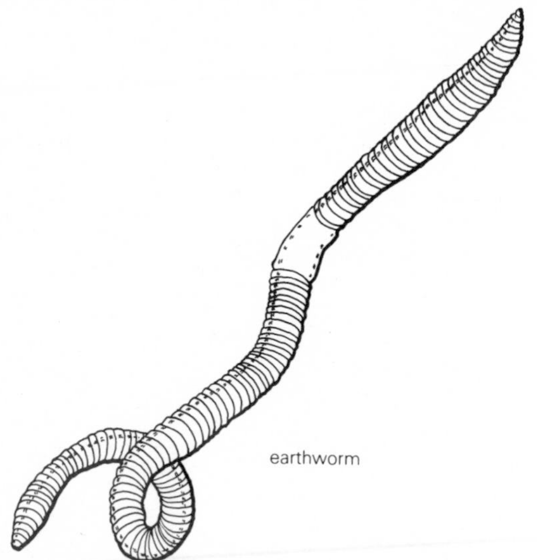

d Using this relationship, explain why a tapeworm does not need a blood system whereas an earthworm does.

68 Intravenous feeding

In hospitals patients are sometimes fed intravenously after operations. To do this, glucose dissolved in a salt solution is allowed to enter a vein slowly.

In an investigation to find the correct strength of salt solution to use, the size of red blood cells was measured, first in a patient's own plasma, and then in a series of salt solutions of different concentrations. The photograph shows red blood cells and two white blood cells in their own plasma. The line drawn on the photograph is 10 micrometres long.

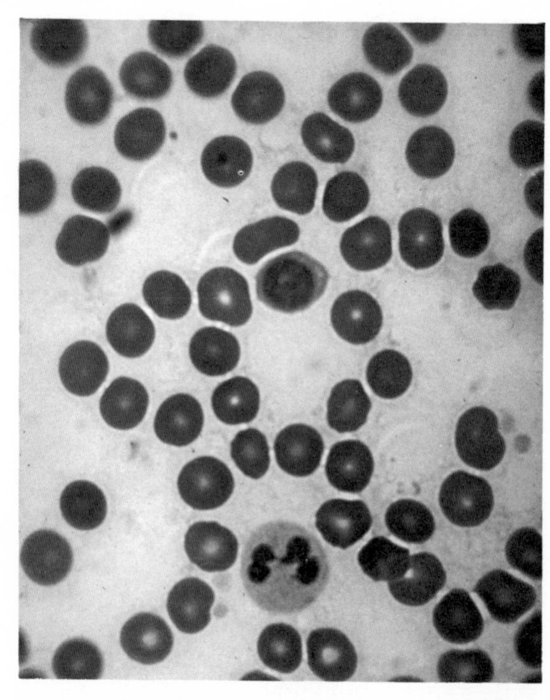

a Estimate the diameter of a red blood cell in its own plasma, by measuring the diameter of several cells and working out the mean diameter.

The graph opposite shows the diameter of red blood cells when placed in different concentratons of salt solution.

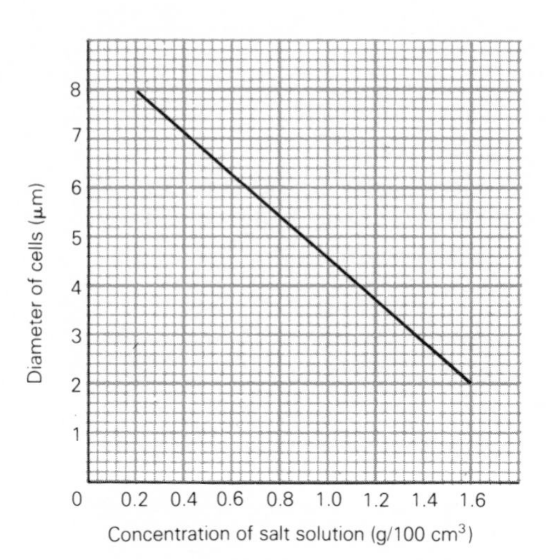

b From the information in the graph, which concentration of salt solution should be used for intravenous feeding? Explain why you chose this concentration.

Many blood donors are now invited to donate plasma rather than whole blood.

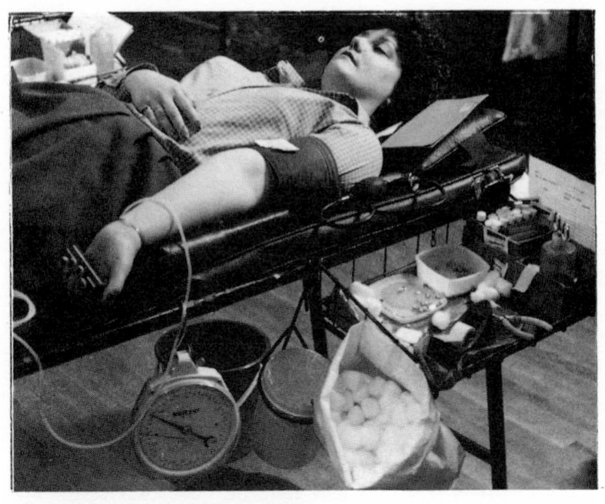

a volunteer donating blood

c Suggest how the plasma might be separated from the blood.

d Explain why a person can donate plasma more frequently than whole blood.

69 Counting red blood cells

Doctors often need to know how many red blood cells a patient has. The number of red blood cells in 1 mm^3 of blood can be calculated by using a special microscope slide like the one shown. A sample of blood is diluted 200 times, e.g. by putting 1 cm^3 in 199 cm^3 of dilute salt solution. A drop of this diluted blood is placed on the special slide, which has four large squares ruled on it. The large squares have a side of 0.2 mm. The depth of the film of liquid is exactly 0.1 mm.

0.2mm

a Explain how you would count the number of cells in one of the large squares.

b Count the number of cells in each of the four large squares and write your results in a suitable table.

c Calculate the mean number of cells in a large square.

d Calculate the number of red cells in 1 mm^3 of this blood.

e A shortage of red blood cells is called anaemia. Anaemia may result from
(i) shortage of iron in the diet,
(ii) shortage of vitamin in B_{12} in the diet – vitamin B_{12} is needed for normal red cell production,
(iii) absence of the substance needed in the intestine for vitamin B_{12} to be absorbed into the blood.
Explain why each of these three factors may cause anaemia, and suggest how each could be treated.

70 Athletes' hearts

Athletic coaches are very interested in the efficiency of athletes' hearts. The table below shows how the output of blood from the heart varies with the number of beats per minute.

a (i) Plot a graph of total output and heart rate. Use the horizontal axis for heart rate, and the vertical axis for total output.
(ii) Plot a graph of output per beat and heart rate. Use the horizontal axis for heart rate and the vertical axis for total output.

b Describe fully the relationship between:
(i) heart rate and total output;
(ii) heart rate and output per beat.

c Successful long distance runners often have low heart beat rates. Explain why this might be of advantage to them.

d No matter how hard we exercise, the heart rate seldom rises above 140 beats per minute. Using the figures above, explain why this is so.

Heart rate (beats/min)	55	70	80	90	120	140	150	170
Total output (litres/min)	4.0	4.8	5.2	5.6	6.0	6.0	5.8	4.6
Output/beat (cm^3)	73	69	65	63	60	43	39	26

71 A baby's blood system

A human foetus receives its food and oxygen from its mother's blood at the placenta. The structure of the foetal blood system is slightly different from an adult's blood system, to help in its functions.

The diagram shows the structure of the foetal blood system. From the diagram you can see that there is a hole in the wall between the auricles of the heart (labelled A), and a vessel (labelled B) connecting the

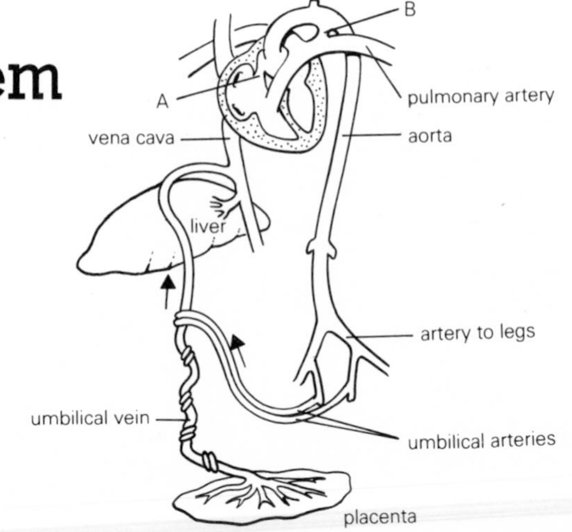

pulmonary artery and the aorta. The hole A and the connection B close just after the birth of the baby.

a Explain why A and B are important to the foetus.

b Suggest what would happen if A and B remained open after birth.

c Describe the most direct route by which oxygen could be carried in the blood from the placenta to the legs of a foetus.

d When the umbilical cord is cut after birth it does not usually bleed, although it contains an artery and a vein. Use your knowledge of the structure of blood vessels to explain how bleeding may be prevented.

72 Heart disease

Operations can often now save the lives of people with heart diseases which used to be fatal. The photograph shows an operating theatre during a heart operation.

a What is the most common cause of blockage in a coronary artery?

b What can a person do to reduce the chances of a coronary artery becoming blocked?

c During an operation to by-pass a blockage in a coronary artery the patient's heart is stopped and kept at a temperature of 10 °C. Suggest why the heart is kept at this temperature.

d The patient's blood is kept circulating by a 'heart–lung' machine as shown in the photograph. Suggest how the blood might be oxygenated in such a machine.

e A substance called heparin which prevents blood clotting is usually added to the blood during the operation. Suggest why this is necessary.

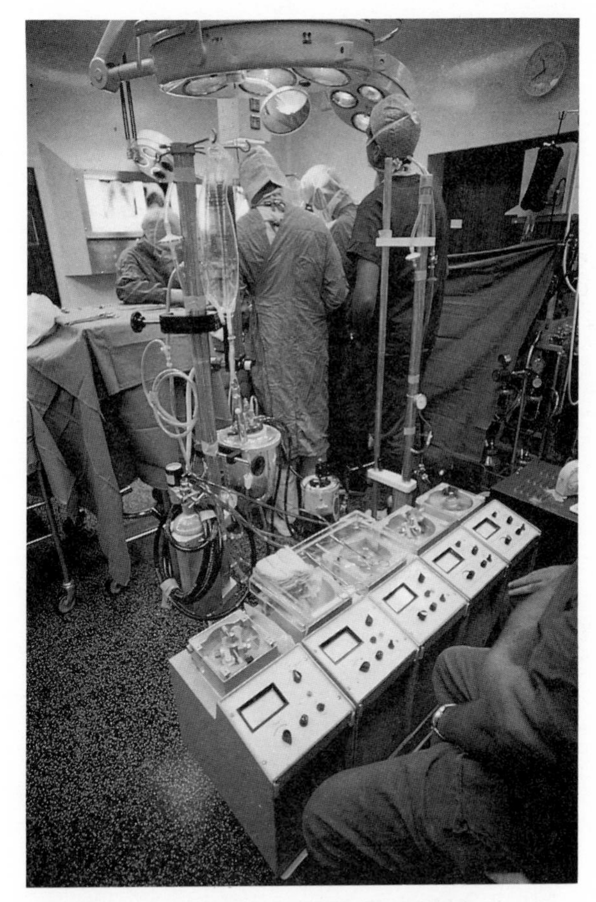

the heart-lung machine is used to maintain the patient's vital functions during the course of the operation

73 Water loss

Water for crops is very scarce in many parts of the world. Scientists are therefore very interested in the rate at which different crop plants lose water to the atmosphere. The apparatus below was used to measure this rate.

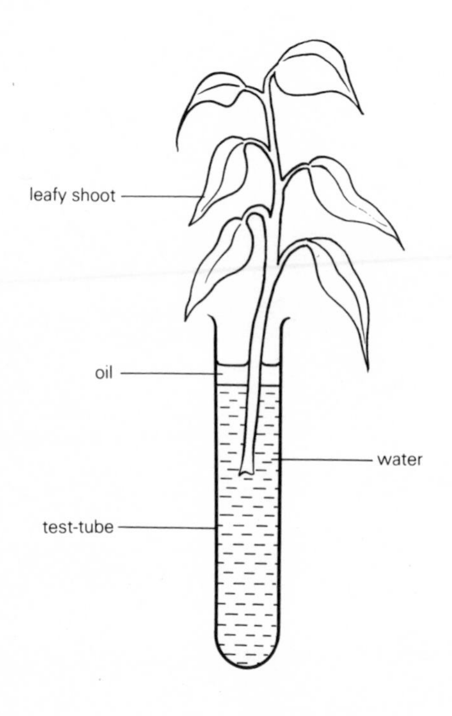

The apparatus was weighed every hour for five hours. It was then put into a large transparent plastic bag and again weighed every hour for five hours. In both investigations, the loss in mass from the start of the experiment was recorded. The results are shown in the table below.

Time (hours)	Loss in mass from the start of the experiment (g)	
	Apparatus in the open	Apparatus in plastic bag
1	0.2	0.1
2	0.4	0.2
3	0.5	0.3
4	0.8	0.3
5	1.0	0.3

a Plot both sets of results on the same graph.

b Explain why it was better to use the same shoot in different conditions rather than to use two different shoots at the same time.

c What was the purpose of putting oil on the water in the tube?

d Explain how water in the tube could be lost to the atmosphere.

e Which conditions were changed by placing the apparatus in a plastic bag?

f Which conditions could have changed during the course of the investigation?

g Besides water loss, two other processes taking place in the plant can affect its mass.
(i) Which other process could cause a decrease in mass?
(ii) Which process could cause an increase in mass?

h From the results of the experiment, suggest how a farmer growing salad vegetables in a hot dry area could conserve water.

74 Prize pears

Some gardeners grow fruit for competitions. Often the gardener with the largest fruit wins the prize. One such gardener noticed that the pears on a branch whose bark had been damaged in early summer seemed much larger than those on undamaged branches. She wondered if taking the bark from the base of a branch would increase the size of her competition pears.

a Design an investigation to find if the fruit grower's suspicion was correct.

b Assuming that the pears did grow larger, suggest an explanation for their increased size.

c Suggest what would happen to the damaged branch the following year.

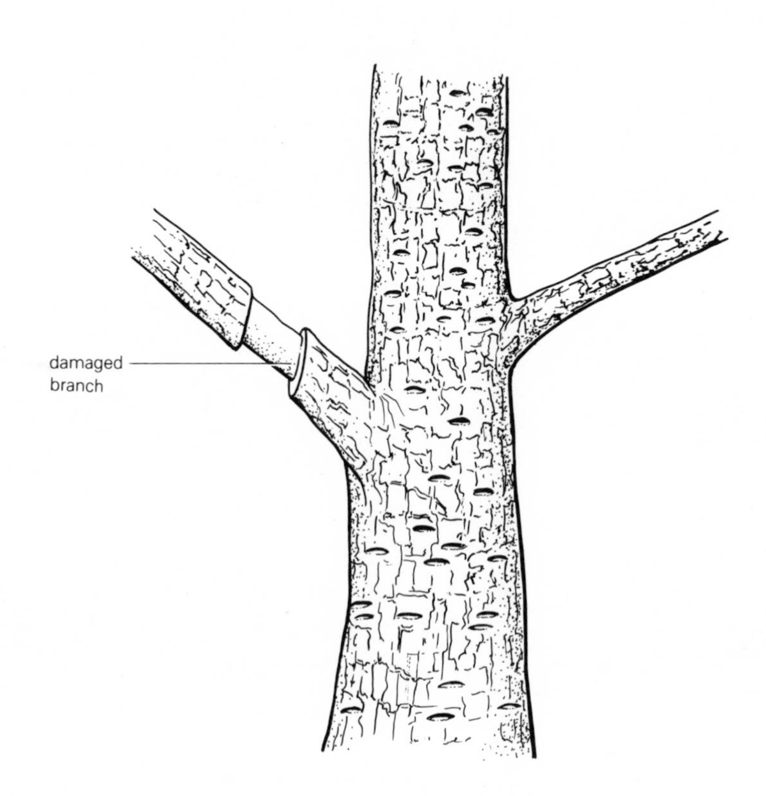

75 Mineral ion transport

Scientists who study the effects of mineral ions on crop growth need to know how these ions are transported through plants. In one investigation to find which tissue in a plant stem transports mineral ions, the xylem and phloem in an experimental shoot were separated by placing waxed paper between them as shown in the diagram overleaf.

The end of the shoot was then placed in a solution containing a radioactive isotope of potassium (^{42}K). After five hours the amount of ^{42}K in various parts was measured. The results are shown in the table overleaf.

a Why was waxed paper used in the experimental shoot?

b What kind of paper, do you think, was used in the control shoot?

c Where was the largest amount of ^{42}K found?

d Where was the lowest amount of ^{42}K found?

e Which tissue is mainly responsible for transporting ^{42}K upwards in the shoot? Explain the reasons for your answer.

	Amount of ^{42}K (parts per million)			
	Shoot with xylem and phloem separated by waxed paper		Control shoot	
	phloem	xylem	phloem	xylem
Above separated section	53	47	64	56
Separated section	0.7	112	87	69
Below separated section	84	58	57	67

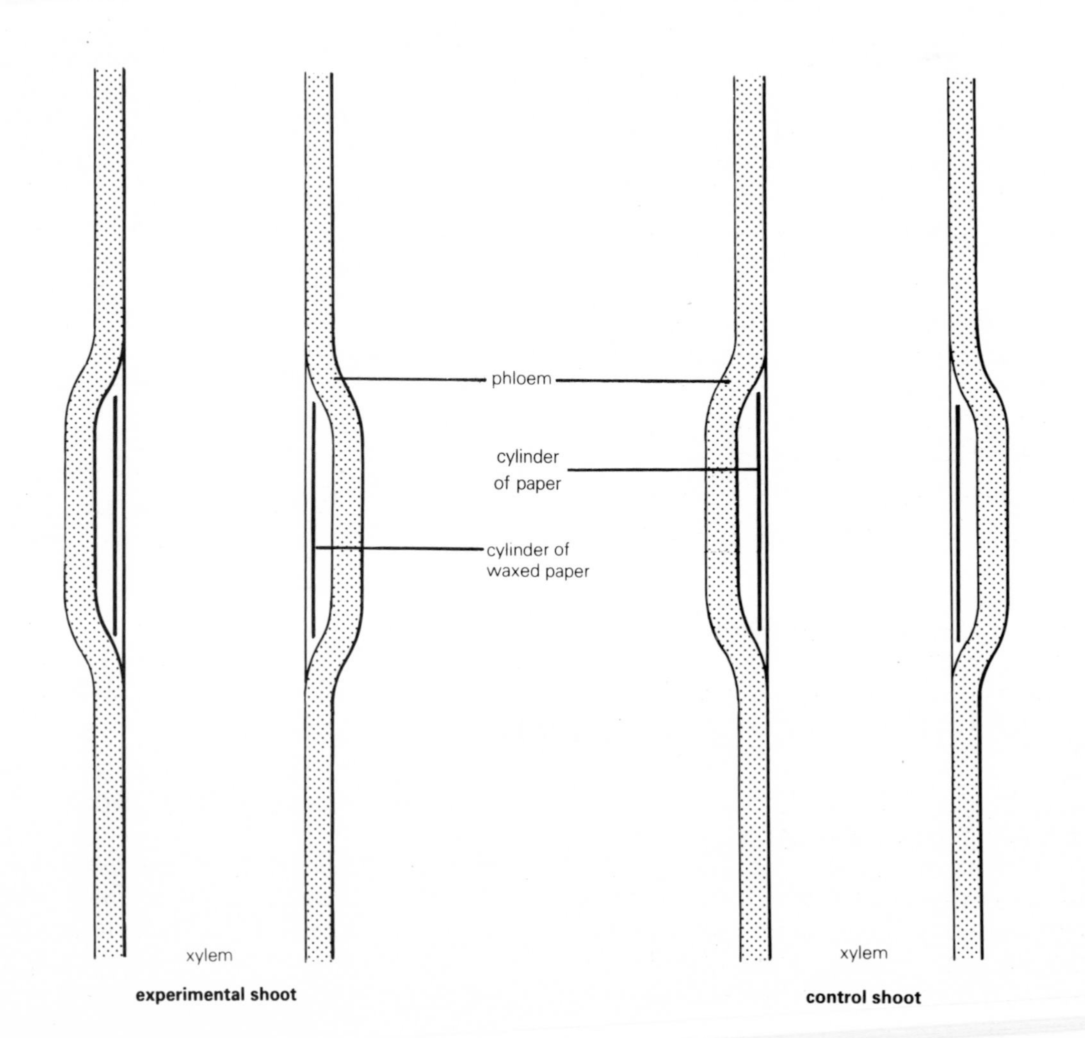

UNIT 10 Respiration

All living organisms need energy, for example for muscle activity, making new substances and maintaining body temperature. Energy is released from food in the process of respiration. To release the maximum amount of energy oxygen is required, and active organisms like mammals need a constant supply of oxygen for respiration. Some energy can be released even without oxygen, and some organisms such as yeast can respire for considerable periods of time without oxygen. The questions in this unit are about how oxygen is obtained and how the energy released in respiration may be used.

76 Exercise

Two boys counted how many times they breathed in half a minute while sitting still. They then did two minutes of vigorous exercise. As soon as they stopped they counted how many times they breathed in each half minute for a total of five minutes. Their results are shown opposite.

Activity	Time scale (minutes)	Number of breaths in each half minute: Andrew	David
Sitting still	0	7	8
		7	8
	1	7	8
		7	8
Exercise	2		
	0		
Sitting still		22	24
	1	18	17
		15	13
	2	13	10
		12	9
	3	12	8
		10	8
	4	8	8
		7	8
	5	7	8

a Draw a graph to show how the rates of breathing of the two boys changed in the five minutes after the vigorous exercise. Use the horizontal axis for time, and plot both lines on the same axes.

b How long did each boy take to return to his normal breathing rate after the exercise?

c Which boy is fitter? Explain your answer.

d (i) What happens to the breathing rate when we exercise?
(ii) Explain the advantages of this change.

77 Breathing

The table overleaf shows some of the results of one of the first experiments carried out to investigate the effect of carbon dioxide concentration on breathing.

Percentage CO_2 in air breathed in	Number of breaths per minute	Total volume of air breathed per minute (litres)
0.04 (normal)	14	9.4
0.8	14	10.3
1.5	15	11.9
2.3	15	13.7
3.1	15	18.5
5.5	16	29.5
6.0	27	56.8

a What happens to the rate of breathing as the carbon dioxide concentration increases?

b What happens to the total volume of air breathed in as the carbon dioxide concentration increases?

c Explain how the total volume of air breathed in per minute can change even though the number of breaths stays the same.

d (i) What will be the effect of increasing carbon dioxide concentration in the air on the ability of the blood to get rid of waste carbon dioxide?
(ii) Explain the advantage of the changes in breathing.

78 Lung structure

The photograph shows a small section across a lung, magnified 100 times. The alveoli are not fully inflated because they collapse when the section is made. They also do not seem to be the same size because they have been cut across at different places and in different directions.

a Calculate the approximate size of an alveolus, by measuring the average internal diameter of five of the larger alveoli.

b (i) If you are good at arithmetic, try working out the internal surface area of an alveolus using the formula for the surface area of a sphere, $4\pi r^2$.
(ii) Explain why the answer you get will not be a totally accurate measure of the internal surface area.

c Suggest how you could estimate the total internal surface area of the lungs.

d In certain diseases some of the walls of the alveoli break down.
(i) What effect will this have on the total surface area of the lungs?
(ii) How will this affect the absorption of oxygen?

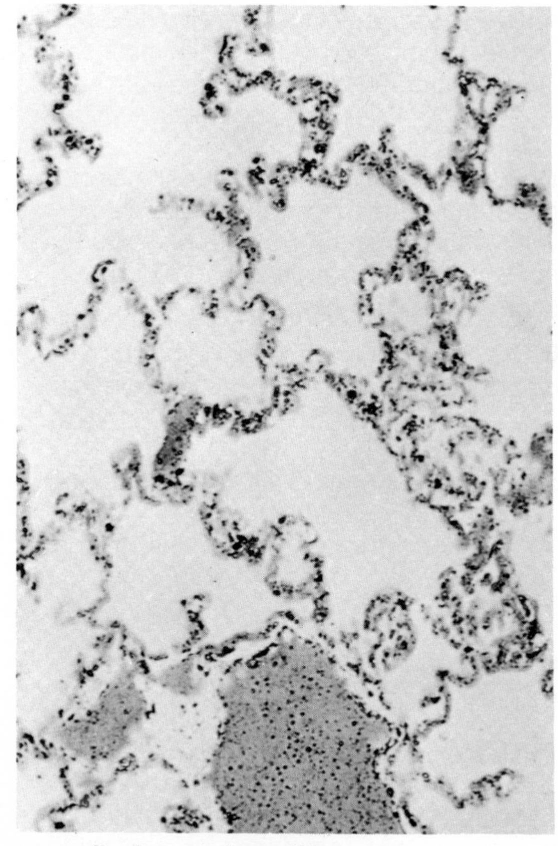

mammalian lung section × 100

The photomicrograph (right) shows a very small part of an alveolus wall, and part of a red blood cell in a capillary alongside it. The magnification is about 150 000 times.

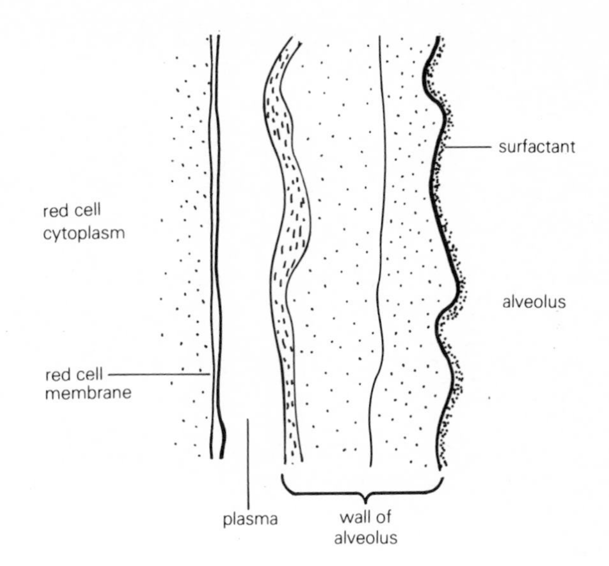

e How far does oxygen have to diffuse to get from the alveolus into the plasma?

f Why must the alveolus wall be so thin?

g Name two substances which would diffuse from the plasma into the cells of the alveolus wall.

h Oxygen diffuses about 10 000 times as fast in air as in water. Explain why you drown if your lungs fill with water.

i The surface layer of the alveoli has a substance, called a surfactant, to stop the alveoli sticking together like an empty plastic bag. What would happen if the surfactant was absent?

79 Lungs and size

The table below shows the body mass of four different mammals. It also shows how much lung surface area they have for every gram of their body mass, and how rapidly they use up oxygen at rest.

a Describe the relationship between body mass and the surface area of the lungs per gram of body mass.

b Suggest an explanation for this relationship, in terms of oxygen consumption.

c Explain why a large mammal like man uses much less oxygen per minute per gram of body mass than a small mammal like a mouse.

d (i) From the data in the table, work out the total surface area of the lungs of a bat.
(ii) Explain how this large surface area is achieved (The lungs are similar in structure to those of other mammals.)

e Suggest why the bat has a larger lung surface area for its size than a mouse, even though its oxygen consumption at rest is less than that of a mouse.

Mammal	Body mass (g)	Surface area of lungs per g of body mass (cm^2)	Volume of oxygen used per minute (cm^3 per g of body mass)
Bat	10	100	1 500
Mouse	20	54	3 600
Rat	300	33	770
Man	70 000	11	200

80 Walking and running

The graph shows the amount of oxygen consumed per minute when a man is walking and running at different speeds on the level.

a What is the rate of oxygen consumption when the man is
(i) walking at 5 km per hour?
(ii) running at 15 km per hour?
(iii) at rest?

b (i) What is the relationship between rate of oxygen consumption and speed of running?
(ii) Suggest an explanation for this relationship.

c List four ways in which the body adapts to allow for the increased oxygen requirement during fast running.

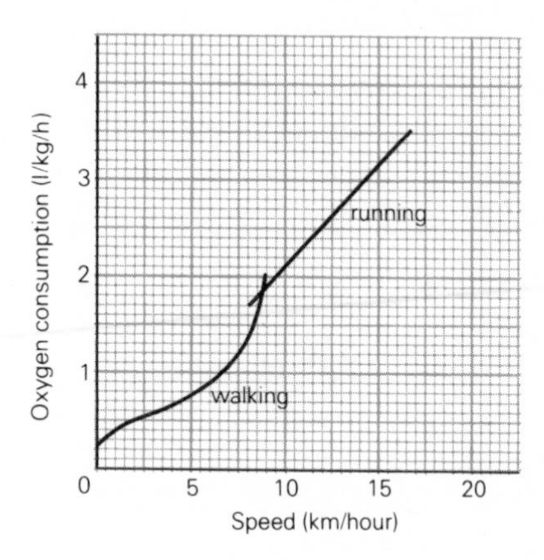

d At what speed does walking become less efficient than running?

e Suggest a reason why fast walking is less efficient than running at the same speed.

81 Running, swimming and flying

A ground squirrel and a salmon both have the same mass of 200 g. The ground squirrel runs along the ground at 3 km per hour, which is the same speed as the salmon swims. The ground squirrel uses 4 kJ of energy to run 1 km. The salmon uses 0.3 kJ to swim 1 km. A bird of similar mass uses about 1.5 kJ to fly 1 km at a speed which is about 20 times faster than the squirrel runs or the salmon swims.

a Which method of movement uses least energy per km?

b How much energy would the salmon use if it swam for 1 hour?

c 1 g of fat provides about 36 kJ of energy. How far could could the bird fly on a store of 1 g of fat?

d How much stored fat would the bird need for a migratory flight of 100 km?

e Is it easier to move through air or water? Give reasons for your answer.

f How is a fish like the salmon adapted to move quickly through water?

g Suggest why the ground squirrel has to use so much energy for running. In your answer consider how energy is wasted in moving over land.

h A migratory bird has to store fat before starting its migration. What factors would determine how much fat must be stored for its migration to be successfully completed?

i Suggest why birds often migrate long distances, but small ground animals rarely migrate far.

82 Diving ducks

The diagram below is copied from a research paper on the effect of diving on the heart rate of a duck. Note that a duck has lungs and can only breathe air; it cannot breathe underwater.

The pictures and block chart show what the duck is doing at each stage:

A swimming on the surface
B preparing to dive
C diving
D feeding on the bottom
E surfacing
F returning to the surface

The ECG line shows a trace of the duck's heartbeat, as recorded electrically. Each vertical line shows one beat of the heart. The bottom line shows the time scale in seconds. Each mark represents one second.

a Describe the main changes in the rate of heartbeat during the stages A to F.

b Use the time scale and ECG trace to calculate the approximate rate of heartbeat in beats per minute during
(i) stage B;
(ii) stage D.

c Suggest why the heart rate increases in stage B.

d What differences would you expect in the breathing rates between stages A, B and C?

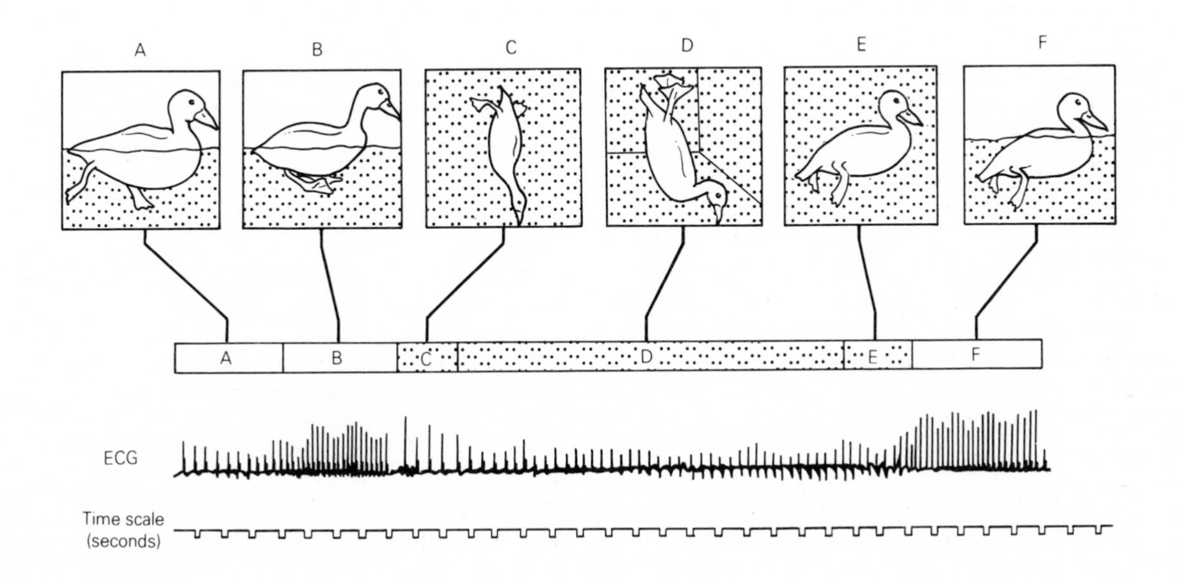

83 Hibernation

Some mammals hibernate for the winter. They become inactive and enter into a 'winter sleep'. Marmots live in the Alps, and in winter they go down into hay-lined burrows where groups of about fifteen huddle together and sleep through the winter. In the table, the body temperature and the rates of breathing and heartbeat of marmots during the summer and winter are compared.

	In summer when active	In winter when hibernating
Average body temperature	37 °C	5 °C
Rate of breathing	30 per minute	1 per 5 minutes
Rate of heartbeat	80 per minute	4 per minute

a What happens to the marmot's body temperature when it hibernates?

b What happens to the the rates of breathing and heartbeat when marmots hibernate?

c What will happen to the rate at which food stores are used up when the body temperature is lowered?

d What will happen to the rate of respiration when body temperature is lowered?

marmot emerging from burrow

e What will happen to the rate at which oxygen is used up when the body temperature is lowered?

f Explain why the rate of breathing changes during hibernation.

g Explain why the rate of heartbeat changes during hibernation.

h The marmots become very fat during autumn, because a thick layer of fat is laid down under the skin. Give two advantages of this layer of fat.

i Suggest why groups of marmots huddle together in hay-lined burrows during winter.

84 Wine

Read the following account of how wine is made.

Ripe grapes are picked and crushed to produce a 'must' which contains juice, pulp, skin and pips. The must is then treated with sulphur dioxide to prevent the growth of unwanted bacteria and yeasts which are present on the grape skins. The juice is fermented by the wine yeast, which also grows on the skin of the grapes. However some wine producers add starter cultures of known strains of the yeast. Complete breakdown of the sugar in the must produces a dry wine with an alcohol content of up to 15%, at which concentration the yeast is killed. Sweet wines are made from musts with very high sugar content, or by stopping fermentation before all the sugar is used up. The fermented liquor is aged in vats, when sediment is deposited. The liquor is then filtered before it is bottled.

a Draw a flow chart to show the process of wine-making, as described in the above account.

Use the account to answer questions b to e.

b Traditional wines are made without adding any yeast. What is the source of yeast in this case?

c What is the purpose of adding sulphur dioxide to the must?

d What concentration of alcohol kills yeast?

e How is a sweet wine made?

f Explain why fermentation occurs only in the absence of oxygen.

g Sparkling wines are made when part of the fermentation takes place in sealed bottles. Explain why the wine produced in this way is 'sparkling'.

h Very ripe, 'late season' grapes usually produce sweet wines. Explain why.

UNIT 11 Keeping a balance

Living cells can only function properly if they have the right conditions. Even small changes in temperature, concentration of solutions or amounts of chemicals can disrupt or even kill the cells. The questions in this unit are about the many different ways living organisms cope with changes in the environment and keep conditions correctly balanced for their cells.

85 Camels in the desert

Some mammals have tissues which can function satisfactorily over a wide range of temperatures. A good example is the camel, which can be active in the full heat of the desert sun. During the day its body temperature gradually increases, reaching 40 °C by sunset (6.00 p.m.). Its temperature then falls to 34 °C by dawn (6.00 a.m.).

a Sketch a graph to show how the temperature of a camel varies over a 24-hour period.

b One reason why its temperature rises during the day is that the camel does not sweat unless its temperature rises above 40 °C. What advantage is it to the camel not to sweat?

c What advantage is it to the camel if its body temperature falls to 34 °C by dawn?

86 A desert plant

Some plants can control their temperature. An example is the California Monkey Flower which lives in the Californian deserts. When the temperature of its leaves reaches 41.5 °C its stomata suddenly open wide. It can then survive in temperatures as high as 60 °C.

a Explain how the opening of the stomata helps to keep the California Monkey Flower cool.

b What is the disadvantage of this method of losing heat?

87 Spiny anteaters

Spiny anteaters are primitive mammals. They have no sweat glands, do not pant and cannot control the blood supply to the surface of the skin.

a (i) Describe how you would expect the blood temperature of the spiny anteater to vary as external temperatures change from 10 °C to 40 °C

spiny anteater

(ii) What would happen to the blood temperature of the human body in the same conditions?

b Suggest how the spiny anteater could avoid overheating on a hot day.

88 Kidney machines

Many people whose kidneys do not function properly can be kept alive by passing their blood through an artificial kidney machine. The patient's blood passes between membranes which let small molecules of water, glucose and urea through, but not large protein molecules or blood cells. Outside the membranes is a solution into which the waste substances diffuse. This form of treatment is called dialysis.

A manufacturer intending to produce the solutions used in artificial kidney machines investigated the composition of blood plasma and urine in healthy humans. Some of the results are shown in the table below.

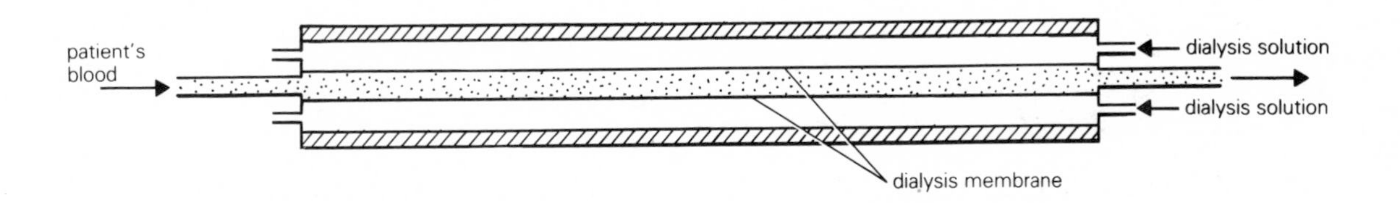

	Plasma concentration (g per 1000 cm^3)	Urine concentration (g per 1000 cm^3)
Water	900–930	950
Protein	70–80	0
Glucose	1.0	0
Urea	0.3	20
Sodium ions	3.2	3.5
Chloride ions	3.7	6

a Why is there no protein in the urine?

b Why is there no glucose in the urine?

c How would the composition of the plasma of a patient requiring dialysis differ from that of a healthy person?

d In the kidney machine, which substance or substances
(i) should be removed from the blood completely?
(ii) should not be removed from the blood?
(iii) should be partially removed to keep the amounts in the blood constant?

e If you were the manufacturer, what would be the composition of the solution you supplied for use in the machine?

89 A crab's problem

A crab lives in rock pools on the seashore. The concentration of salts in these pools can vary considerably.

a Suggest one factor which could increase the concentration of salts in a rock pool.

b Suggest one factor which could decrease the concentration of salts in a rock pool.

c Explain how a decrease in concentration of salts would affect the crab.

A crab was placed in solutions of different salt concentrations and its oxygen consumption was measured. The results obtained are shown in the graph.

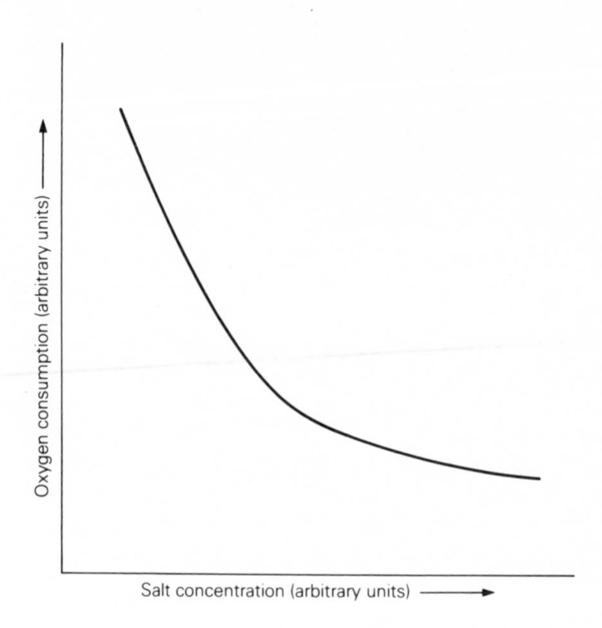

d What happens to the oxygen consumption as the salt concentration falls?

e Suggest why the oxygen consumption changes as the salt concentration falls.

90 Diabetics

A diabetic person does not produce enough of the hormone insulin to keep the blood sugar concentration at a low enough level after meals. One way of testing if a person has diabetes is to give a glucose tolerance test. After not eating for twelve hours, the patient is asked to drink a glucose solution containing 50 g of glucose. The amount of glucose in the blood is then measured every 30 minutes for the next three hours. The results of giving this test to a diabetic patient and the same test to a healthy person are shown in the graphs opposite.

a Why should the people being tested not eat for twelve hours before the test?

b What was the greatest concentration of glucose in the diabetic's blood?

c How long did it take for the glucose concentration of the healthy person to return to the starting level?

d Estimate from the graph how long it would take for the glucose concentration of the diabetic to return to the starting level.

e What effect would injecting insulin into the diabetic have on the results of the test?

f Why do diabetics have to inject insulin every day rather than once a week?

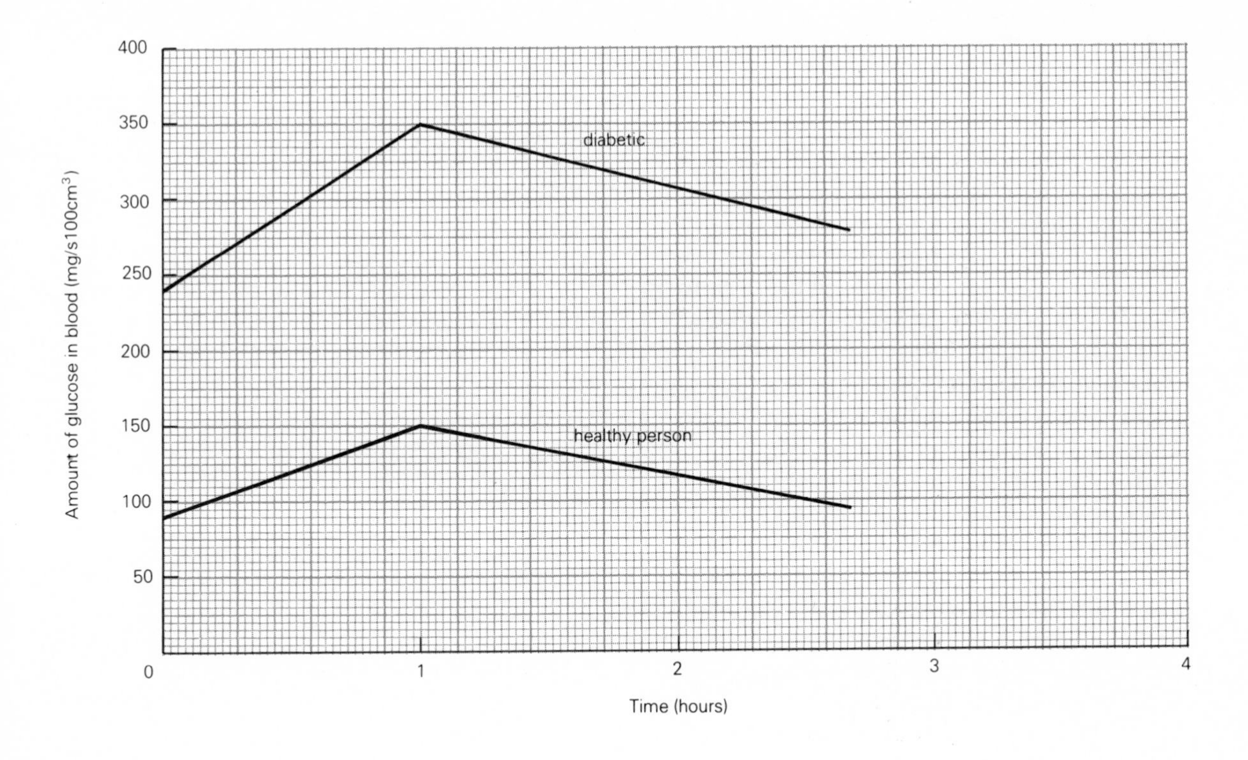

91 An aquatic toad

Xenopus is a toad which normally lives in water. Its two main excretory products are ammonia and urea. In an investigation a toad was taken out of the water for four days and then put back into the water. The amounts of urea and ammonia it excreted under different conditions were measured and are shown in the table.

a Plot a bar chart of the results.

b Explain why the volumes of ammonia and urea excreted changed when the toad was out of water.

c While the toad was out of water it continued to make ammonia. Explain what happened to this ammonia.

	The day before removal from the water	Three days after removal from the water	The day after return to the water
Volume of ammonia excreted (arbitrary units)	25	5	30
Volume of urea excreted (arbitrary units)	15	4	110

Control and co-ordination

The body of a complex organism like a mammal depends on a large number of different activities working together in an organised way. For example, when a human being runs, many muscles have to contract and relax in a precisely co-ordinated order, the rates of breathing and heartbeat have to be increased, extra sugar may be needed from the liver, and the excess heat may have to be lost from the skin. All these activities are controlled by a combination of the nervous system and chemical hormones. Even in less active organisms co-ordination is essential. Organisms also have to respond to changes in their surroundings in order to obtain food, move away from danger, keep cool and so on. The questions in this unit are about how organisms control their activities and respond to changes.

92 Reaction times

The apparatus shown in the diagram below was used by pupils in a biology class to test how quickly each pupil could respond to a traffic light.

When the start switch is pressed by one pupil, the red light comes on, and the subject has to press the 'brake pedal' as quickly as possible. The timer measures the pupil's reaction time in hundredths of a second. The class results are shown in the table.

Pupil	Reaction time (hundredths of a second)	Pupil	Reaction time (hundredths of a second)
1	35	16	19
2	27	17	34
3	15	18	18
4	27	19	27
5	24	20	19
6	29	21	23
7	23	22	22
8	24	23	28
9	26	24	16
10	28	25	21
11	18	26	20
12	32	27	23
13	25	28	25
14	37	29	31
15	31	30	24

a Complete a table like the one on the next page to show how many pupils there are with various reaction times.

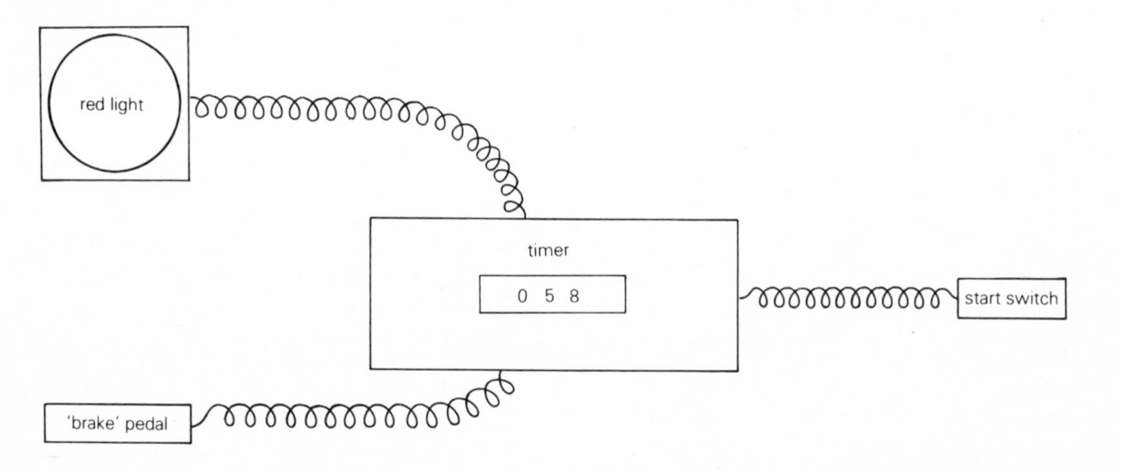

Reaction time (hundredths of a second)	15–19	20–24	25–29	30–34	35–39
Number of pupils					

b Draw a histogram of the results.

c Which pupil had
(i) the fastest reaction?
(ii) the slowest reaction?

d Calculate the mean reaction time for the pupils in the class.

e Calculate the percentage of pupils with a reaction time greater than 24 hundredths of a second.

f If a car was being driven at 60 km per hour, how far would the car go before
(i) the pupil with the fastest reaction time hit the brake pedal?
(ii) the pupil with the slowest reaction time hit the brake pedal?

93 Eyes

As people get older the lenses in their eyes lose their ability to change shape. In a young person the lens becomes thicker and more convex as the tension in the lens is reduced when the ciliary muscles contract. In an older person the lens stays thin.

a Explain why it is an advantage for the lens to be able to change shape and become thicker.

The graph shows the average maximum thickness that the lens can reach at different ages.

b (i) At what age does the lens more or less lose the ability to change shape?
(ii) What effect will this have on the person's ability to see?

c What change in ability to see will occur between the ages of 10 and 20?

d What should be the shape of the lenses in the glasses older people use for reading? Explain your answer.

e From the information in the graph, explain why people over 60 do not have to keep getting new glasses for reading.

f In some old people the lens in the eye becomes so cloudy that it has to be removed in an operation. Explain how it is still possible for the eye to form an image even with the lens removed.

g Plan an investigation to find out the greatest distance at which a person can see an object which is only 2 mm in size.

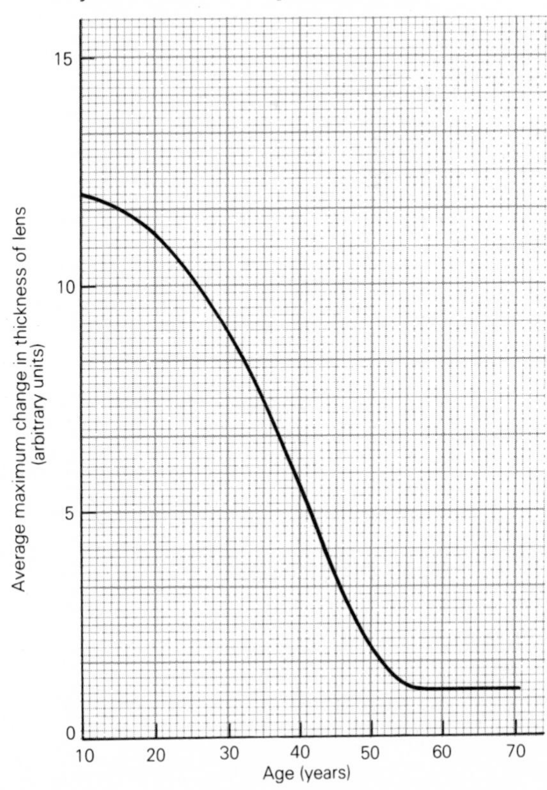

94 The thyroid gland

The thyroid gland produces a hormone, called thyroxine, which affects the rate of metabolism. If the thyroid gland does not produce enough thyroxine in an adult human, the person has a very low metabolic rate and is said to have myxoedema; if the thyroid gland produces too much thyroxine, the person has a very high metabolic rate and is said to have hyperthyroidism. Thyroxine contains iodine, and iodine is absorbed by the thyroid gland to make thyroxine.

A radioactive isotope of iodine, ^{131}I, can be used to measure the activity of the thyroid gland. The patient is given a drink containing a small amount of the radioactive isotope; the amount of radioactive iodine in the thyroid gland is measured over the next two days. The graph shows the results of testing a healthy person, a person with myxoedema and a person with hyperthyroidism.

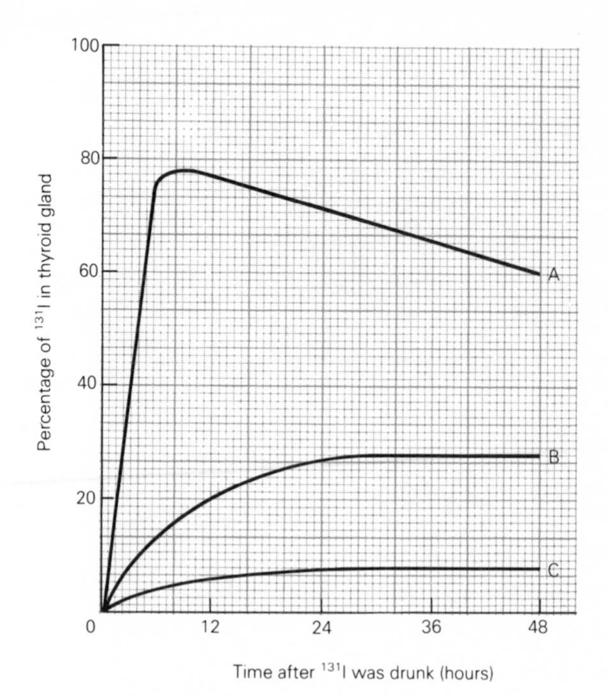

a Which line on the graph shows the patient with myxoedema?

b Which line on the graph shows the patient with hyperthyroidism?

c Suggest what effects myxoedema would have on a patient.

d Suggest how the amount of the radioactive isotope in the thyroid gland could be measured.

e Suggest three properties which a radioactive isotope should have to make it suitable and safe for use in an investigation like the one described.

f An accident at a nuclear power station could release large amounts of the radioactive isotope of iodine. In the event of such an accident some people might receive large doses of radioactive iodine. To protect them, the people at risk would be given tablets containing non-radioactive iodine. Suggest how these tablets could help to protect them.

95 Phototropism

Plant shoots grow towards sunlight, a response called phototropism. In an investigation to find which colour of light is most effective in bringing about this response, the following results were obtained. The chart opposite shows the relationship between wavelength and colour.

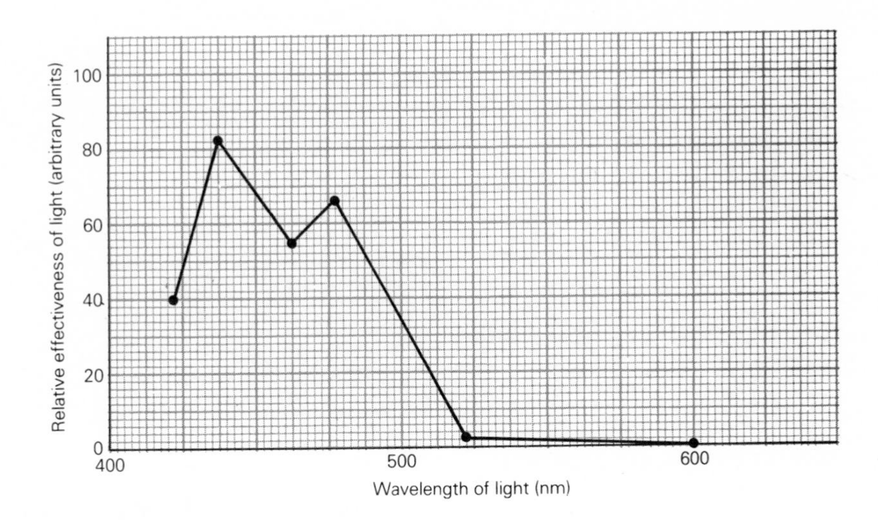

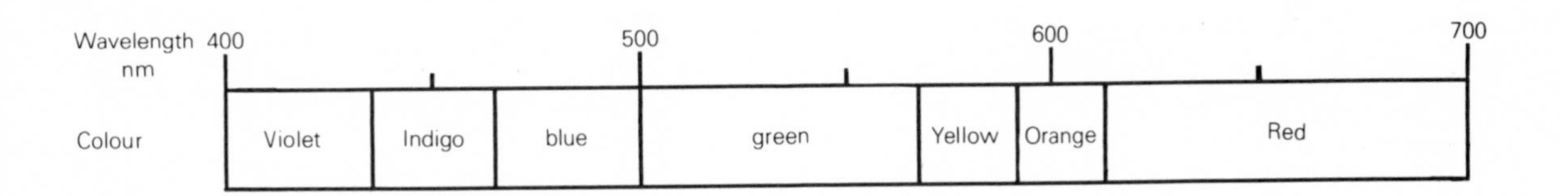

a (i) Which wavelength of light is most effective in bringing about phototropism?
(ii) What colour is this light?

b (i) What would be the effect of shining white light onto a shoot from one side?
(ii) What would be the effect of shining red light onto a shoot from one side?

c Explain the advantage of the phototropic response to a plant.

d Suggest how the investigation described above could have been carried out, using straight young shoots. Explain how you could measure the 'effectiveness' of different colours of light.

96 Geotropism

Geotropism is a plant's response to gravity. Cress seeds may be germinated by placing them between horizontal sheets of damp blotting paper. If the top sheet of paper is removed when the seedlings have grown, the shoots soon grow upwards. If, however, the seedlings are left in darkness they seem to reach the upright position much more quickly than when left in the light.

a Design an investigation to find if the light does affect the speed of this geotropic response.

b Suggest a reason why the seedlings left in light respond more slowly to gravity than those left in darkness.

97 Sticklebacks

In the breeding season the male stickleback builds a shallow, covered nest at the bottom of a pond. After attracting the female to lay eggs in the nest, he sheds his sperm over the eggs to fertilise them. He remains near the nest until the eggs hatch, occasionally fanning his tail to drive a current of water over the eggs.

a Suggest an advantage of driving a current of water over the eggs.

The time spent by a male fanning his tail was measured as the eggs developed. The results are shown in the graph.

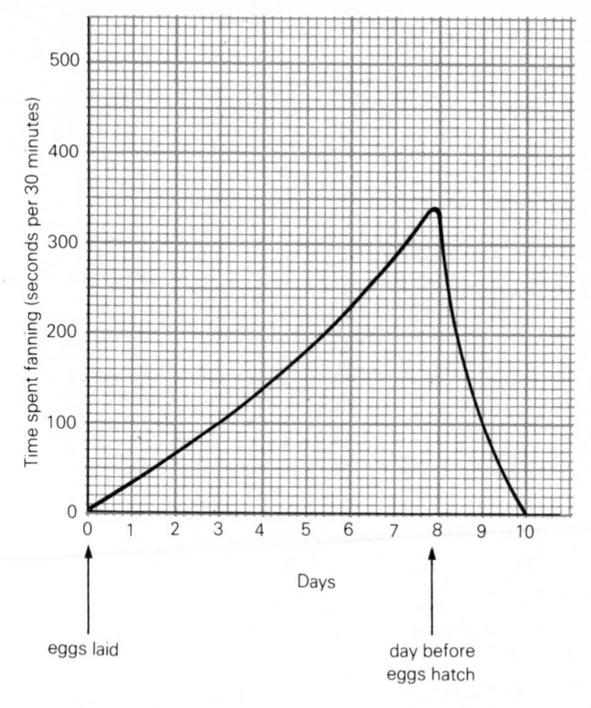

b On which day is most time spent fanning?

c Suggest what stimulates the male to increase the frequency of fanning as the eggs develop.

d In a second investigation, the original clutch of eggs was taken away on day six and replaced with a clutch of newly laid and fertilised eggs.

(i) On which day would you expect these eggs to hatch?
(ii) Sketch a graph to show the time that you would expect the male to spend on fanning, up to the time of hatching of the second clutch of eggs.

98 Waterfleas

Waterfleas are tiny crustaceans which live in freshwater ponds. They are often fed to goldfish in aquaria. When the waterfleas are placed in an aquarium they swim horizontally until they pass underneath floating plants; they then swim upwards towards the plants.

a Waterfleas cannot eat the plants. Suggest two possible reasons why the waterfleas swim towards the plants.

b In an aquarium water plants do not cast a shadow, since the sides of the aquarium are transparent. The waterfleas cannot 'see' the plants as they do not have eyes which can form images. Suggest how they find the plants.

UNIT 13 Growth and reproduction

The survival of a species depends on its ability to reproduce itself. Reproduction may simply involve separating into two parts, or splitting off part of the organism. Most organisms, however, have the ability to reproduce sexually by producing special cells which can join together in fertilisation. The great advantage of sexual reproduction is that by mixing features together it allows the offspring to be different from their parents and possibly better suited to conditions. Whichever method of reproduction is used the offspring have to grow to reach adult size. In many-celled organisms this involves repeated cell division. The questions in this unit are about the processes involved in growth and reproduction.

99 Growing up

The graph shows how much an average boy gains in mass per year from birth to the age of 18.

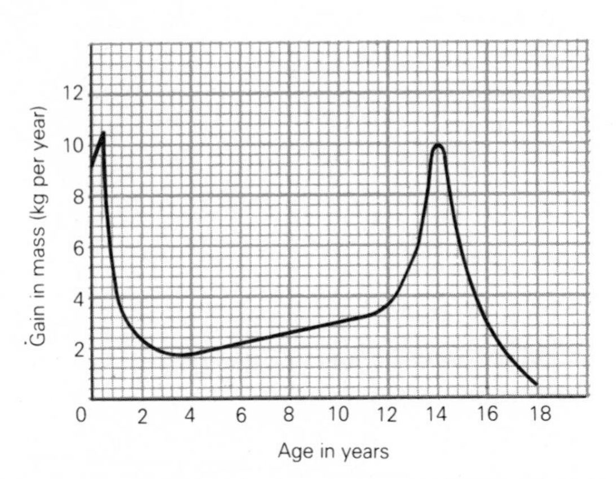

a Approximately how much mass is gained between the years of
(i) 4 and 5?
(ii) 14 and 15?

b At what age is the rate of growth fastest?

c Sketch a graph to show the overall change in mass of average boys up to the age of 18.

d Suggest *three* factors which might affect the rate of increase in mass for an individual boy.

e A graph showing gain in height shows a similar pattern to that for mass. How would you expect the two graphs to differ if they were continued beyond the age of 18?

Very small amounts of some elements can affect the growth and development of embryos and of children. The histograms overleaf show the amounts of three different elements found in the placental tissues of babies at birth, compared with the average mass of the baby at birth.

f Describe the effect of increasing concentration of each of the three elements on the mass of the baby at birth.

g Which of the three elements are likely to be damaging development in the concentrations found?

h Consider the following information.
(i) Lead is contained in car exhaust fumes because lead is added to petrol.
(ii) The concentration of cadmium in placental tissues is increased as a result of smoking.
(iii) High concentrations of cadmium and lead slow down the uptake of zinc by an embryo.

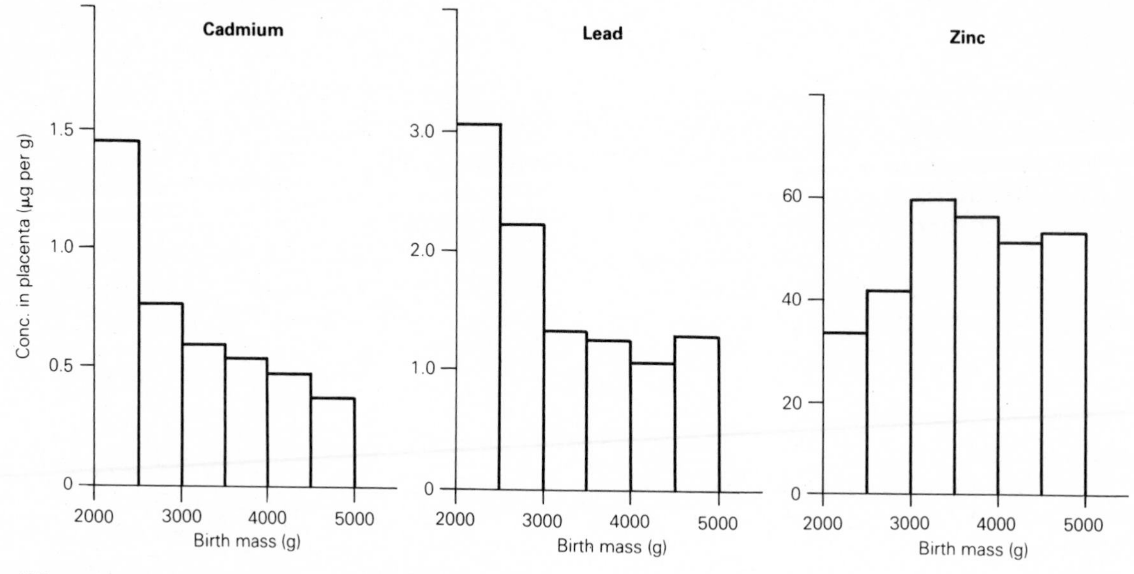

(*Note*: 1 microgram (μg) is one millionth of a gram.)

What conclusions about possible hazards to healthy development would you draw from this information?

100 Ageing

The table below shows the mean times taken by men of different ages to complete a 30 km cross-country race in Sweden.

Age group (years)	Mean time (min)
16–20	158
21–25	157
26–30	155
31–35	157
36–40	162
41–45	167
46–50	172
51–55	178
56–60	178
61–65	210

a Plot a histogram to show these results.

b By which age group are the best performances recorded?

c (i) Which result appears not to follow the general pattern?
(ii) Suggest a reason for this odd result.

d Suggest two changes in the body which might explain why the time taken to complete the race increases with age.

e The average life span of human beings in Europe has increased considerably during this century. The maximum life span is probably about 120 years. List four problems which would arise if most people lived to be 120.

101 Cell division

The photograph shows some of the cells from the active growing region in the tip of a root. In some cells the nucleus looks like a dark circle (e.g. cell A). In others, the thread-like chromosomes can be seen (e.g. Cells B and C). The cells in which the chromosomes can be seen are in the process of dividing.

a Count how many of the cells were dividing when the photograph was taken.

b Work out the percentage of the cells that were dividing.

c (i) Approximately what proportion of a non-dividing cell is filled by the nucleus?
(ii) How does this compare with a cell such as a palisade cell from a leaf?

d Use a textbook to find diagrams of the various stages of cell division (mitosis) and list the main stages.

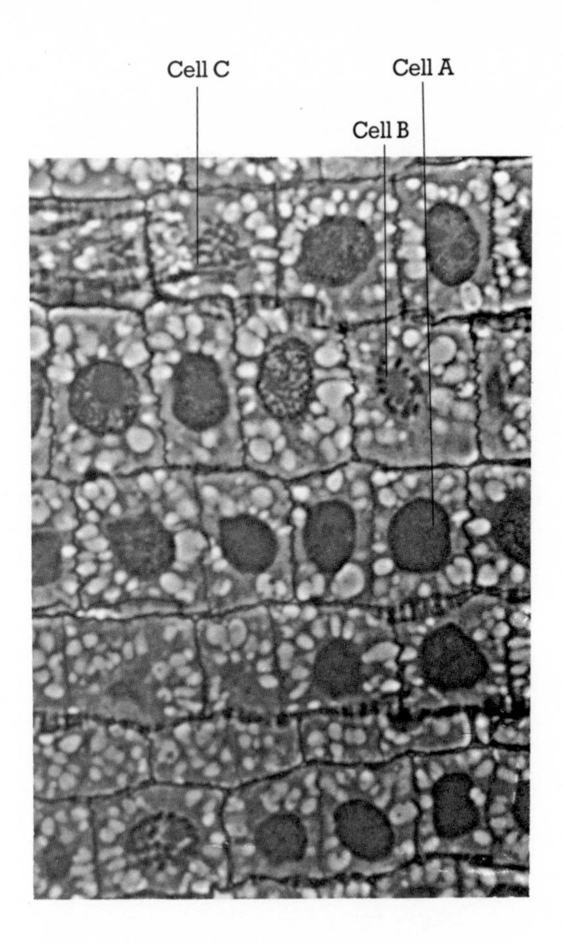

102 Flowers

There are two types of primrose flowers, known as pin-eyed and thrum-eyed. Both types are shown in the diagram below. Pin-eyed and thrum-eyed flowers grow together in the same area, but pollen from one type of flower can only fertilise egg cells in the other type of flower.

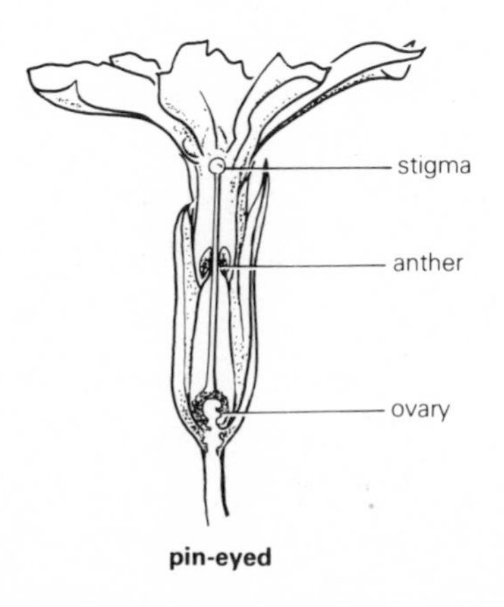

pin-eyed

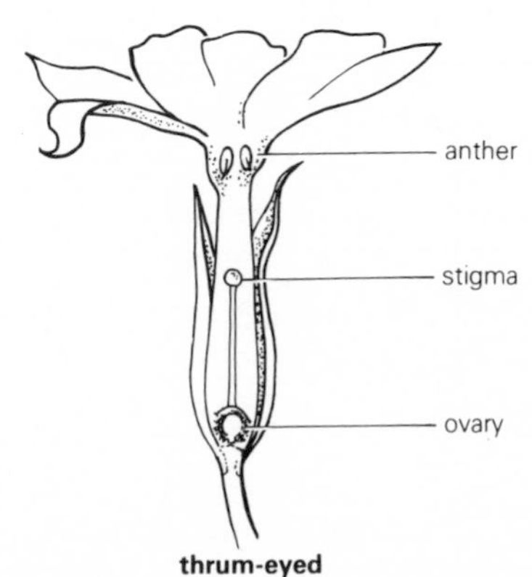

thrum-eyed

Primrose

a Describe two differences between the pin-eyed flower and the thrum-eyed flower.

b Describe two pieces of evidence, which you can see from the diagrams, to suggest that the primrose is insect-pollinated.

c Primroses may be pollinated by a bee which can extend a long tongue into the flower. Explain how a bee would pollinate a thrum-eyed flower with pollen from a pin-eyed flower.

d Explain the biological advantage of this system of pollination.

The red campion also has two types of flower. The diagrams below show sections through each type.

e What are the main differences between the two types of red campion flower?

f Describe how the red campion's method of ensuring cross-pollination is different from that of the primrose.

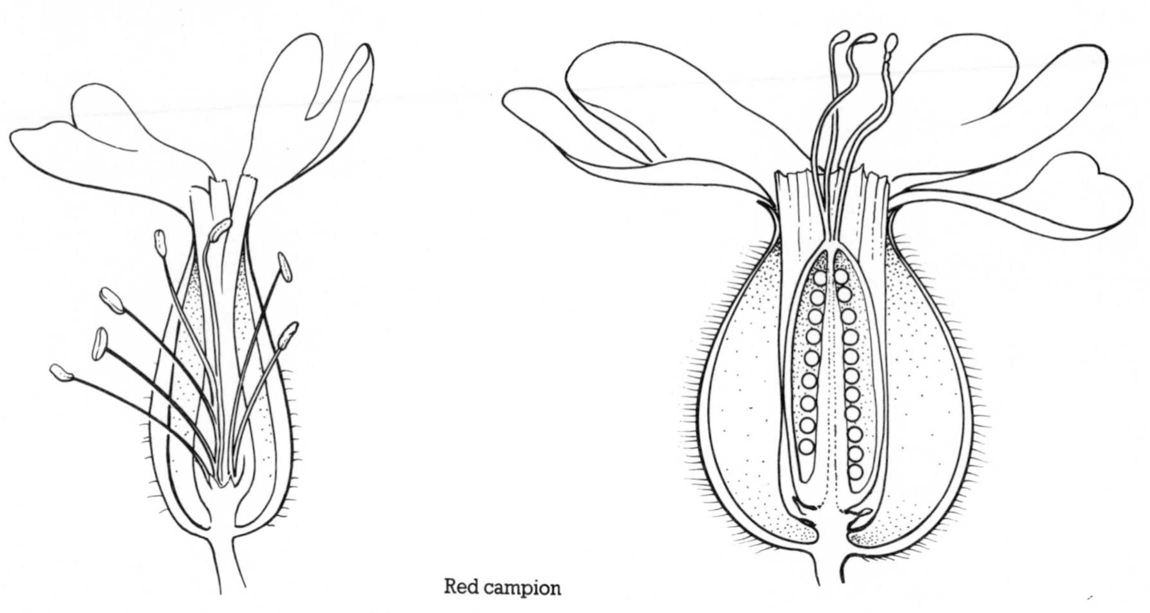

Red campion

103 Human chromosomes

The chromosomes in human cells can be seen by a special treatment in which the chromosomes are made to shrink and are then stained. The photograph shows all the chromosomes in a single nucleus which have been treated in this way.

a Count the total number of chromosomes in this cell.

b (i) Bearing in mind that chromosomes are paired in adult cells, what do you notice about the number of chromosomes in this cell?
(ii) Suggest an explanation for this number of chromosomes being present.

c The X and Y chromosomes have been labelled.
(i) Is the cell from a male or a female?
(ii) Describe the difference between the X and Y chromosomes.

d The pattern of banding enables geneticists to work out the homologous

pairs. The six largest chromosomes (i.e. the largest three pairs) have been labelled A to F. Which of the labelled chromosomes should be paired together?
(Note that the pattern of bands is not identical in each pair, but the general position and arrangements are similar. A really difficult observation exercise is to work out *all* the pairings!)

104 Population growth

The size of a human population depends on:

(i) the birth rate;
(ii) the death rate;
(iii) immigration and/or emigration.

The number of people at different ages in a population can tell us whether the population of that country is growing, falling or staying the same.

A population pyramid for the United Kingdom, which has a stable population, is shown below.

Using the same key, draw the age pyramids you would expect for:

a India (whose population is growing rapidly);

b a rich European country with a falling population;

c an oil-rich Middle Eastern country which has attracted many immigrant workers.

105 Contraception

Contraception is one way of controlling the size of human populations. The pregnancy rate is measured by finding the number of women who get pregnant when 100 women use a method of contraception for one year. This can be used to compare the reliability of different methods of contraception. The table below shows one set of estimates.

Method	Pregnancy rate (per 100 women per year)
Cap and spermicide	3
Contraceptive pill	less than 1
Intra-uterine device (IUD)	3
Safe period	7–15
Condom alone	12
Condom and spermicide	3
Spermicide alone	25
Withdrawal	more than 30

a Which of these methods is the most reliable?

b Which of these methods is the least reliable?

c What is the effect of using spermicide as well as a condom?

d Suggest why a range of figures is given for the safe period method.

e What other factors might be taken into account when deciding which method of contraception to use?

106 Gestation

The table shows how many days it is between fertilisation and birth (i.e. the gestation time) for some mammals.

Mammal	Gestation time (days)	Number of young per litter	Number of litters per year	Mass of adult (kg)
Badger	180	3 to 5	1	15
Cat	60	3 to 6	2	4
Chimpanzee	270	1	1	75
Elephant	640	1	1 per 2 yrs	7000
Guinea pig	60	2 to 6	2 to 3	0.8
Hedgehog	60	3 to 7	1 to 2	0.8
Horse	335	1 to 2	1	1300
Mouse	21	4 to 8	4 to 6	0.025
Pig	115	6 to 20	2 to 3	300
Rabbit	30	4 to 10	3 to 4	1.5
Rat	22	5 to 14	2 to 7	0.5
Whale (Blue)	330	1	1 per 2 yrs	120 000

a Rearrange these mammals in order of mass, starting with the smallest. Write the gestation time alongside each.

b Explain, with reasons, whether each of the following statements is supported by the evidence in the table.
(i) The larger the animal, the longer the gestation time is.
(ii) Mammals of the same mass have the same gestation time.
(iii) The smaller the mammal, the more young there are in a litter.
(iv) The number of litters per year varies according to the length of gestation tme.
(v) Large mammals have only one young per litter.

c Suggest three factors which might have an effect on the gestation time of a particular species of mammal.

d Estimate the approximate gestation time for the following mammals:
(i) the hare (mass 5 kg), which has 2 to 5 young in a litter, and 2 to 4 litters per year;
(ii) the lion (mass 250 kg), which has 2 to 4 young once a year.

e The average gestation time in human beings varies according to the number of babies born, as you can see from the table below.

Number of babies born	Average gestation time (days)	Average mass of each baby (kg)
Single baby	280	3.4
Twin	262	2.4
Triplets	247	1.8
Quads	237	1.4

(i) What is the total mass of the babies born in each case?
(ii) Suggest an explanation for the variation in gestation times.

107 Eggs

Below is some information on the eggs of various animals.

Animal	Average number of eggs produced at one time	Size (mm)
Human	1 to 2	0.2
Mouse	8	0.1
Swan	5 to 7	115
Sparrow	3 to 5	22
Turtle	100	40
Frog	2 000	2
Cod	5 000 000	0.1

a In which of these animals is fertilisation internal?

b In which of these animals do the young develop internally?

c Suggest why some eggs are so much larger than others.

d Suggest why some animals produce such a large number of eggs.

e If the number of cod in the sea remains more or less the same, on average how many of the cod's eggs will survive to produce a mature adult fish? Explain your answer.

Inheritance and evolution

When living organisms reproduce they pass on some of their characteristics to their young. The small differences between members of a species can give some individuals an advantage, which helps them to survive and leads to a process of gradual change or evolution. The questions in this unit are about the processes by which characteristics are passed on from one generation to the next and about how the resulting variations can lead to change in a species.

108 Fingerprints

It is said that no two people have identical fingerprints. The patterns of fingerprints, however, can be classified into three main types – arches, loops and whorls. Examples of these are shown in the photographs. You will also see a photograph of a fingerprint found at the scene of a crime, and of twelve people whose prints were taken as suspects.

a List the type of fingerprint (arch, loop or whorl) which each suspect has.

b Who committed the crime?

Among the suspects are two pairs of identical twins. They are John and Peter, and Michael and Graham. Robert and Howard, and Mary and Louise are also twins, but are not identical.

c If a fingerprint were entirely the result of a person's genes, what results would you expect when you examined the twins' fingerprints?

d Does each pair of twins have similar fingerprints?

e What do these results tell you about the inheritance and development of fingerprints?

ARCH

LOOP

WHORL

print found at the
scene of the crime

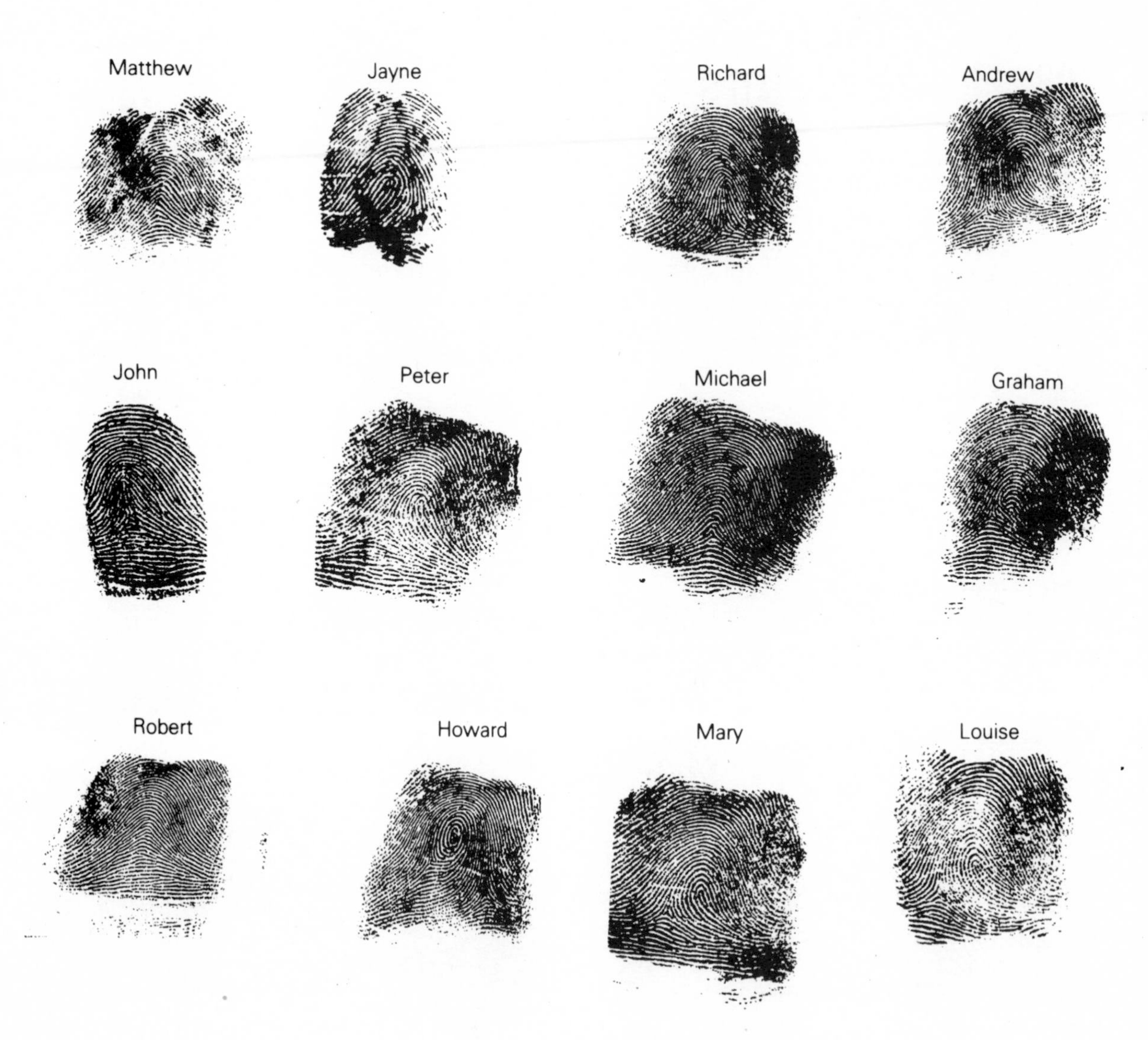
Matthew
Jayne
Richard
Andrew
John
Peter
Michael
Graham
Robert
Howard
Mary
Louise

109 Tobacco plants

The photograph shows a tray of tobacco seedlings, all grown from the seeds of one parent plant. Some of the seedlings appear very pale in the photograph because their leaves contain no chlorophyll. These seedlings are called albino. The seedlings with dark leaves do have chlorophyll and are normal.

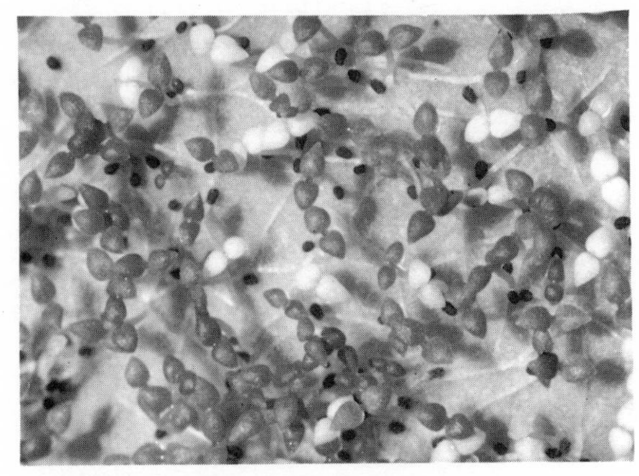

normal and albino tobacco seedlings

a Count the numbers of normal seedlings and albino seedlings. Record your results in a suitable table.

b What is the approximate ratio of normal: albino seedlings?

Chlorophyll production is controlled by a gene. The allele which produces chlorophyll may be written as *C*. The absence of chlorophyll is caused by a recessive allele, *c*.

c What would be the genotypes of the albino seedlings?

d What must the genotypes of the parent plants have been? Explain your answer with a genetic diagram.

e Approximately how many of the normal seedlings in the tray would you expect to be homozygous (i.e. to have both alleles the same)?

f Explain why the albino seedlings would not produce seeds.

110 Maize

Maize cobs ('corn on the cob') have a large number of grains arranged in rows. Each grain is produced from a different ovule and so can be genetically different. In the photograph the grains have either purple or yellow coats. The colour is the result of one pair of alleles. The maize cob in the photograph was the result of crossing two purple-grained maize plants.

a Which grain colour is dominant?

b Count the number of purple grains and the number of yellow grains in each of the central four rows. Present your results in a table.

c What is the ratio of purple grains to yellow grains?

d Draw a genetic diagram to show the cross which produced the cob in the photograph.

purple and yellow seeded maize cob

111 Oil palms

The fruits of the oil palm are an important source of vegetable oil, used in many products such as margarine. The fruits of the wild oil palm from Africa have thick shells and a fairly thin fleshy outer layer. The vegetable oil comes from this fleshy layer. A few palms produce fruits with no shell, but only the pollen can be used to produce seeds as the female parts of the flower are sterile. Hybrid plants, produced by crossing thick-shelled and shell-less palms, have thin-walled shells, as you can see from the photograph.

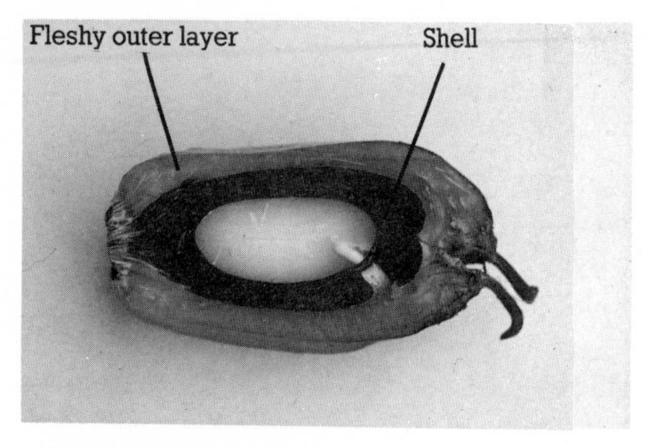

thick shelled oil palm fruit

a Thick-shelled and shell-less palms are both genetically homozygous. The allele responsible for the thick shell is given the symbol, *S*. Absence of shell results from having the allele, *s*. What would be the genotypes of
(i) a thick-shelled palm?
(ii) a shell-less palm?
(iii) a thin-shelled hybrid?

b Explain how hybrid palms would be obtained.

Thick-shelled
Shell-less
Thin-shelled

thin shelled fruit from hybrid plants

c What advantages would the hybrid palm have for the grower?

d If two hybrid palms were crossed, what proportion of the young palms would produce fruits with thin shells?

e Until a few years ago it was not possible to produce clones of the hybrid palms (that is, to reproduce them asexually). Now hybrid palms have been grown from tissue cultures, so the young hybrid palms are genetically identical to their parents. Explain why it is much more economical to grow plants produced from clones than to grow plants produced by cross-breeding the hybrid palms.

112 Blood groups

As well as the well-known A, B, AB and O blood group system, there are also other human blood groups. One of these is the M, N system. Red blood cells may either carry the antigen M, or the antigen N, or both.

The blood group results from the

presence of a single pair of alleles, *M* and *N*. These alleles show no dominance, so if an individual is heterozygous he or she will have the blood group MN. If only *M* or only *N* alleles are present, he or she will have blood group M or N respectively. This information is summarised in the table below.

Blood group	Antigens present	Genotype
M	M only	*MM*
N	N only	*NN*
NN	Both M and N	*MN*

In an investigation into the inheritance of this blood group, a large number of families were tested. The results of these tests are shown below.

Blood groups of parents	Number of families	Blood groups of children		
		M	MN	N
M × M	24	98	1	0
N × N	6	0	0	27
M × N	30	0	43	1
M × MN	86	183	196	0
N × MN	71	0	156	167
MN × MN	69	71	141	63

Two of the results of this investigation are unexpected or anomalous.

a Which are the anomalous results?

b Explain why you would not have expected these results.

c Assuming that the blood tests were correctly carried out, suggest **two** possible explanations for these results.

d What ratios of blood groups in the children would you expect from each of the following crosses?
(i) M × MN
(ii) N × MN
(iii) MN × MN
In each case draw a genetic diagram to show how you worked out your answers.

e (i) Calculate the actual ratios obtained for each of the crosses you have worked out in part **d**.
(ii) Do these results support the genetic theory? Explain your answer.

f A woman, whose blood group is N, has a child whose blood group is also N. Her partner, whose blood group is MN, claims that the child is not his.
(i) What advice would you give on the basis of this evidence?
(ii) How would you confirm whether the child really was the partner's?

113 Ladybirds

The two-spot ladybird is a small beetle. There are two varieties: one which is mostly bright red with two black spots, and one which is mostly black.

The ladybirds feed on small insects such as greenfly. They are eaten by some birds, but many birds do not eat ladybirds because they have an unpleasant taste; the red colour acts as a warning.

Ladybirds hibernate during the winter, and the chart overleaf shows the percentage of the black variety in spring and in autumn over a period of four years. The percentage in spring shows the proportion emerging from hibernation, the percentage in autumn shows the proportion just before hibernation.

a What was the percentage of black ladybirds in spring of year 1?

b What was the percentage of red ladybirds in spring of year 1?

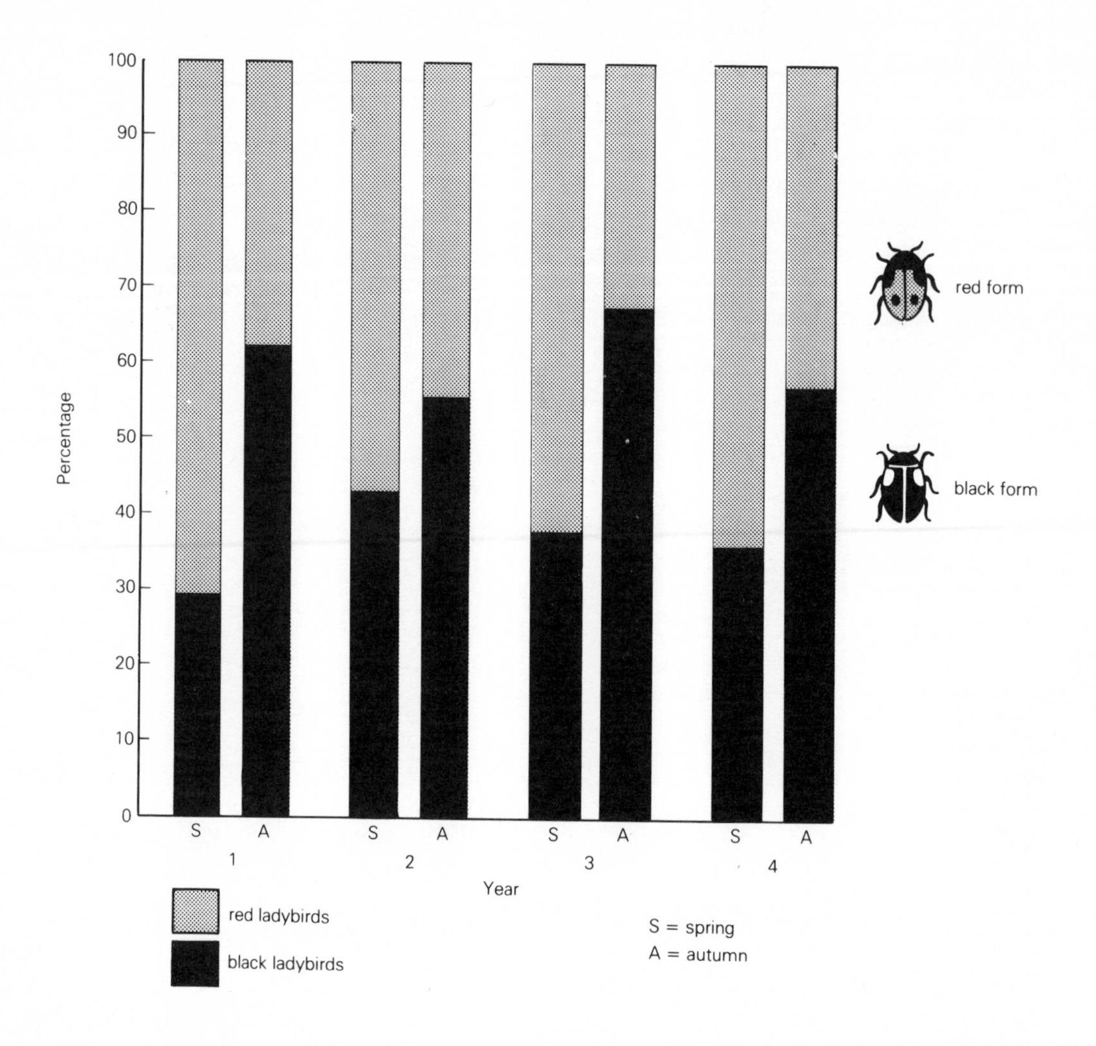

c What was the percentage of black ladybirds in spring of year 3?

d What was the percentage of red ladybirds in autumn of year 3?

e What was the pattern of change in the percentage of the black variety in each of the four years?

f In which year was there the smallest change in proportion during the summer?

g (i) Which coloured variety is more successful at surviving hibernation?
(ii) Suggest a possible explanation for this.

h Suggest two possible causes of the change in proportion of red and black ladybirds that occurs during the summer.

i The black colour is caused by a dominant allele, and only one pair of alleles is involved in determining colour. Draw genetic diagrams to show what young could be produced if a black ladybird mated with a red ladybird. Use *B* to represent the allele for black colour, and *b* for red.

j What proportions of black and red ladybirds could be produced from each mating?

k Suggest why the red variety of ladybird does not become extinct.

114 The peppered moth

Three forms of the peppered moth are found in Britain.

The map (overleaf) shows the results of collecting the moths at 113 different places.

The circles on the map show the proportion of each form of moth caught at each collection point, e.g.

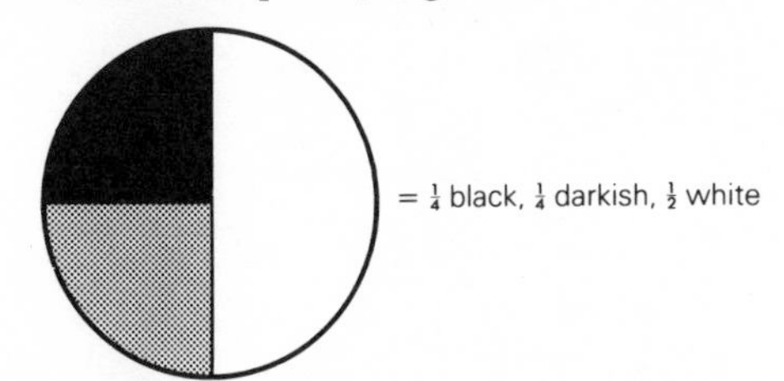

The larger the circle, the more moths that were caught.

a For each of the following, say whether from these results it is:

A certainly true;

B possibly true, but there is not enough evidence;

C untrue.

In each case give your reasons.

(i) In none of the collections were all the moths black.

(ii) At no collection point were the darkish moths the commonest.

(iii) Darkish moths occur only where there are also black and white moths.

(iv) Black moths do not occur in Devon and Cornwall.

(v) There are very few peppered moths in Scotland.

(vi) Black moths are commonest in the coldest parts of Britain.

(vii) Black moths are commoner than white moths in the East of England.

(viii) Black moths are commoner than white moths in industrial areas.

(ix) The peppered moth is rare in industrial areas.

(x) Black moths are better camouflaged than white ones in industrial areas.

b In collections of peppered moths made over 100 years ago, black moths were very rare in all parts of Britain. They only became common in the second half of last century. What conclusions can you draw from this and the information above?

In an experiment, black moths and white moths were released together in two different woods. One wood was in a polluted area near Birmingham. The other was an unpolluted wood in Dorset. The moths were marked with spots of paints before release. In polluted areas the trees have dark trunks and no lichens. In unpolluted areas the trees are covered with whitish lichens. The peppered moths rest during the day on tree trunks.

After a couple of days the moths in the two woods were caught by attracting them to a light. The table (overleaf) shows the percentage recaptured.

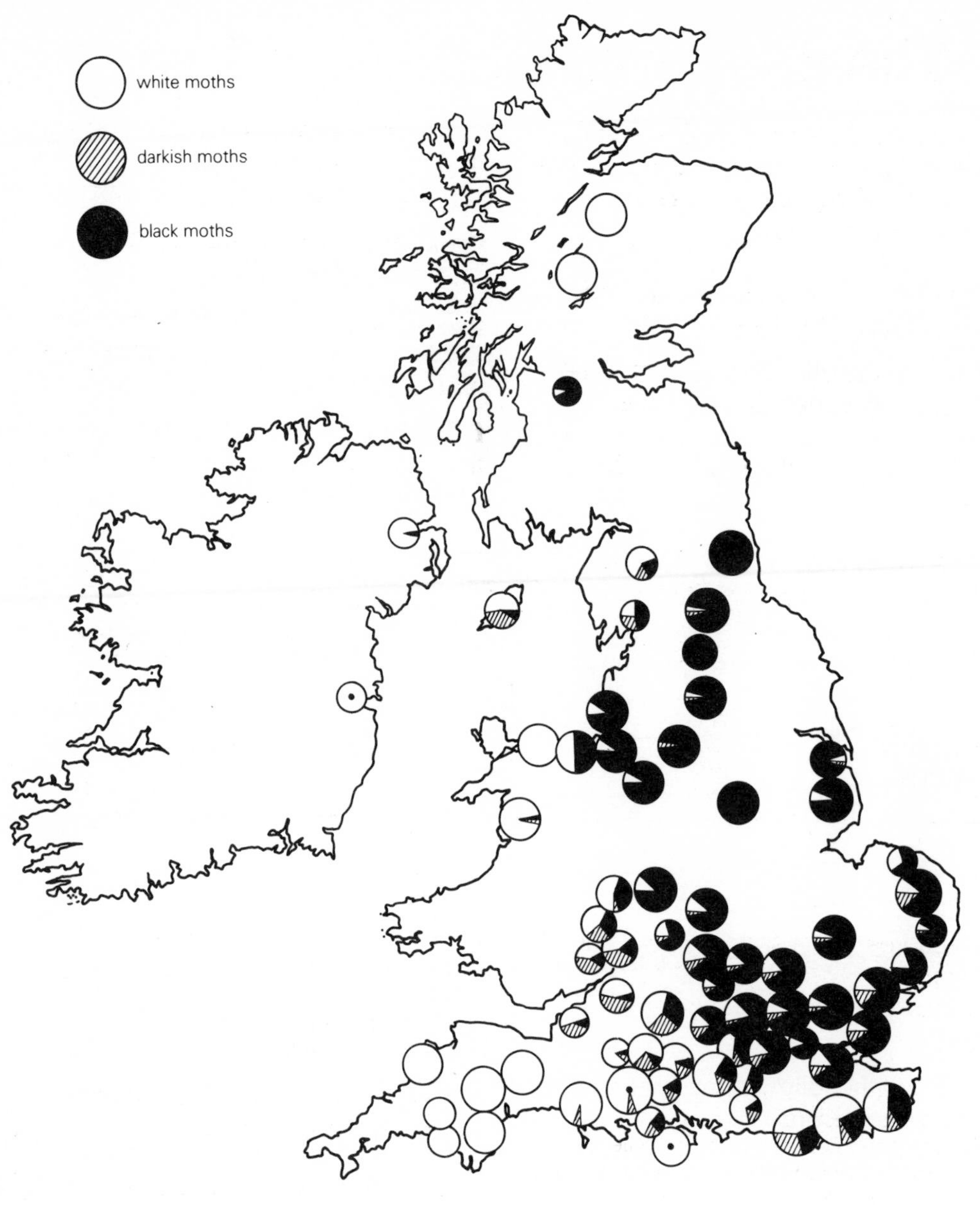

	% recaptured	
	White	Black
Birmingham wood	16	34
Dorset wood	12	6

c (i) Of which form of moth were more recaptured in Birmingham?
(ii) Of which form of moth were more recaptured in Dorset?

d (i) Which form survived better in the polluted wood?
(ii) Which form survived better in the unpolluted wood?

e Suggest an explanation for these results.

f Suggest **two** possible reasons why a higher percentage overall of moths was recaptured in Birmingham than in Dorset.

g Where should the spots of paint have been put when marking the moths? Why?

115 Wheat

The table below shows some information about five different varieties of wheat. These varieties have been artificially bred at various times during this century.

a Plot bar charts to show the grain yields and total biomass of the five varieties.

b (i) Describe the change in height of the wheat as the new varieties have been developed.
(ii) Suggest two advantages of this change in height.

c As new varieties have been developed,
(i) how has the grain yield changed?
(ii) how has the total biomass changed?

d Explain what is meant by the 'total biomass'.

e Explain the advantage to the wheat grower of the relative changes in grain yield and total biomass.

f Describe three factors which could affect the yield actually obtained by a farmer from a field of wheat.

Variety	Year variety was first grown	Height (cm) (length of stem)	Grain yield (tonnes per hectare)	Total biomass (tonnes per hectare)
Little Joss	1908	142	6.0	16.5
Holdfast	1935	126	6.0	16.5
Capella Desprez	1953	110	6.7	15.9
Maris Huntsman	1972	106	7.5	16.3
Norman	1980	84	8.7	17.1

The following paragraph highlights some of the problems involved in growing wheat. Read the passage and then answer the questions.

> 'There can be few more unnatural plant communities than a field of wheat. In medieval times, such a field would have comprised a mixture of locally adapted "land varieties", each showing a reasonable adaptation to losses due to pests and diseases. But modern wheat crops normally comprise pure stands of intensively bred high yielding varieties, and it is not unusual to find a single variety occupying large areas in the main wheat growing districts. Such monocultures provide an ideal environment for the rapid spread of plant pathogens (disease-causing organisms) specifically adapted to such commonly grown varieties. It is therefore not surprising that the history of wheat cultivation in Britain and elsewhere has been characterised by the expansion and decline of a succession of varieties, each lasting a few years until devastated by new strains of disease.'

g What is the disadvantage of growing only one variety of wheat over a large area?

h Why do plant breeders have to keep developing new varieties of wheat?

i The seeds of natural and artificially bred varieties of wheat are preserved in 'seed banks', so that future generations of plant breeders can grow them. Explain why it is important to keep such seed banks, even though we have a number of good varieties of wheat today.

j A plant breeder has a variety of wheat which gives a good yield, but is affected by a fungal disease. Other varieties of wheat are not affected by the same fungal disease. Explain how he could try to breed a variety with the same good yield, but which would not be affected by the fungal disease.

116 Adaptation

Biologists presume that as a result of evolution by natural selection every species of plant and animal has features which adapt it for survival. In many cases it is clear what the survival value of a particular feature is. Often, however, it is difficult to know what the advantage of a feature either was or is. In this case we have to guess. Sometimes it is possible to test a suggested explanation, but often, without being able to go back in time, there is no way of being certain how or why a particular adaptation first arose.

Suggest hypotheses to explain the possible advantage of each of the following observations. In some cases there may be an answer of which we can be fairly certain; in others there is no recognised 'right' explanation.

a Some people have blue eyes, others brown.

b Zebras have stripes.

c Robins have red breasts.

d Earthworms come to the surface when the ground is tapped.

e Dogs turn round several times before sitting down.

f Rabbits have fluffy white tails.

g Herring gulls have a red spot on their bills.

h Oak trees shed all their leaves each autumn.

i Moths fly towards light.